Reviews of Environmental
Contamination and Toxicology

VOLUME 174

T0135098

Springer

New York
Berlin
Heidelberg
Barcelona
Hong Kong
London
Milan
Paris
Singapore
Tokyo

Reviews of Environmental Contamination and Toxicology

Continuation of Residue Reviews

Editor
George W. Ware

VOLUME 174

Springer

Coordinating Board of Editors

Springer-Verlag
New York: 175 Fifth Avenue, New York, NY 10010, USA
Heidelberg: Postfach 10 52 80, 69042 Heidelberg, Germany
ISBN 978-1-4419-2968-6

Printed in the United States of America.

ISSN 0179-5953

Printed on acid-free paper.

Printed in the United States of America

www.springer-ny.com

Springer-Verlag   New York Berlin Heidelberg
A member of BertelsmannSpringer Science+Business Media GmbH

Foreword

International concern in scientific, industrial, and governmental communities over traces of xenobiotics in foods and in both abiotic and biotic environments has justified the present triumvirate of specialized publications in this field: comprehensive reviews, rapidly published research papers and progress reports, and archival documentations. These three international publications are integrated and scheduled to provide the coherency essential for nonduplicative and current progress in a field as dynamic and complex as environmental contamination and toxicology. This series is reserved exclusively for the diversified literature on "toxic" chemicals in our food, our feeds, our homes, recreational and working surroundings, our domestic animals, our wildlife and ourselves. Tremendous efforts worldwide have been mobilized to evaluate the nature, presence, magnitude, fate, and toxicology of the chemicals loosed upon the earth. Among the sequelae of this broad new emphasis is an undeniable need for an articulated set of authoritative publications, where one can find the latest important world literature produced by these emerging areas of science together with documentation of pertinent ancillary legislation.

Research directors and legislative or administrative advisers do not have the time to scan the escalating number of technical publications that may contain articles important to current responsibility. Rather, these individuals need the background provided by detailed reviews and the assurance that the latest information is made available to them, all with minimal literature searching. Similarly, the scientist assigned or attracted to a new problem is required to glean all literature pertinent to the task, to publish new developments or important new experimental details quickly, to inform others of findings that might alter their own efforts, and eventually to publish all his/her supporting data and conclusions for archival purposes.

In the fields of environmental contamination and toxicology, the sum of these concerns and responsibilities is decisively addressed by the uniform, encompassing, and timely publication format of the Springer-Verlag (Heidelberg and New York) triumvirate:

Reviews of Environmental Contamination and Toxicology [Vol. 1 through 97 (1962–1986) as Residue Reviews] for detailed review articles concerned with any aspects of chemical contaminants, including pesticides, in the total environment with toxicological considerations and consequences.

Bulletin of Environmental Contamination and Toxicology (Vol. 1 in 1966) for rapid publication of short reports of significant advances and discoveries in the fields of air, soil, water, and food contamination and pollution as well as methodology and other disciplines concerned with the introduction, presence, and effects of toxicants in the total environment.

Archives of Environmental Contamination and Toxicology (Vol.1 in 1973) for important complete articles emphasizing and describing original experimental or theoretical research work pertaining to the scientific aspects of chemical contaminants in the environment.

Manuscripts for *Reviews* and the *Archives* are in identical formats and are peer reviewed by scientists in the field for adequacy and value; manuscripts for the *Bulletin* are also reviewed, but are published by photo-offset from camera-ready copy to provide the latest results with minimum delay. The individual editors of these three publications comprise the joint Coordinating Board of Editors with referral within the Board of manuscripts submitted to one publication but deemed by major emphasis or length more suitable for one of the others.

Coordinating Board of Editors

Preface

Thanks to our news media, today's lay person may be familiar with such environmental topics as ozone depletion, global warming, greenhouse effect, nuclear and toxic waste disposal, massive marine oil spills, acid rain resulting from atmospheric SO_2 and NO_x, contamination of the marine commons, deforestation, radioactive leaks from nuclear power generators, free chlorine and CFC (chlorofluorocarbon) effects on the ozone layer, mad cow disease, pesticide residues in foods, green chemistry or green technology, volatile organic compounds (VOCs), hormone- or endocrine-disrupting chemicals, declining sperm counts, and immune system suppression by pesticides, just to cite a few. Some of the more current, and perhaps less familiar, additions include *xenobiotic transport, solute transport, Tiers 1 and 2, USEPA to cabinet status, and zero-discharge*. These are only the most prevalent topics of national interest. In more localized settings, residents are faced with leaking underground fuel tanks, movement of nitrates and industrial solvents into groundwater, air pollution and "stay-indoors" alerts in our major cities, radon seepage into homes, poor indoor air quality, chemical spills from overturned railroad tank cars, suspected health effects from living near high-voltage transmission lines, and food contamination by "flesh-eating" bacteria and other fungal or bacterial toxins.

It should then come as no surprise that the '90s generation is the first of mankind to have become afflicted with *chemophobia*, the pervasive and acute fear of chemicals.

There is abundant evidence, however, that virtually all organic chemicals are degraded or dissipated in our not-so-fragile environment, despite efforts by environmental ethicists and the media to persuade us otherwise. However, for most scientists involved in environmental contaminant reduction, there is indeed room for improvement in all spheres.

Environmentalism is the newest global political force, resulting in the emergence of multi-national consortia to control pollution and the evolution of the environmental ethic. Will the new politics of the 21st century be a consortium of technologists and environmentalists or a progressive confrontation? These matters are of genuine concern to governmental agencies and legislative bodies around the world, for many serious chemical incidents have resulted from accidents and improper use.

For those who make the decisions about how our planet is managed, there is an ongoing need for continual surveillance and intelligent controls to avoid endangering the environment, the public health, and wildlife. Ensuring safety-in-use of the many chemicals involved in our highly industrialized culture is a dynamic challenge, for the old, established materials are continually being displaced by newly developed molecules more acceptable to federal and state regulatory agencies, public health officials, and environmentalists.

Adequate safety-in-use evaluations of all chemicals persistent in our air, foodstuffs, and drinking water are not simple matters, and they incorporate the judgments of many individuals highly trained in a variety of complex biological, chemical, food technological, medical, pharmacological, and toxicological disciplines.

Reviews of Environmental Contamination and Toxicology continues to serve as an integrating factor both in focusing attention on those matters requiring further study and in collating for variously trained readers current knowledge in specific important areas involved with chemical contaminants in the total environment. Previous volumes of *Reviews* illustrate these objectives.

Because manuscripts are published in the order in which they are received in final form, it may seem that some important aspects of analytical chemistry, bioaccumulation, biochemistry, human and animal medicine, legislation, pharmacology, physiology, regulation, and toxicology have been neglected at times. However, these apparent omissions are recognized, and pertinent manuscripts are in preparation. The field is so very large and the interests in it are so varied that the Editor and the Editorial Board earnestly solicit authors and suggestions of underrepresented topics to make this international book series yet more useful and worthwhile.

Reviews of Environmental Contamination and Toxicology attempts to provide concise, critical reviews of timely advances, philosophy, and significant areas of accomplished or needed endeavor in the total field of xenobiotics in any segment of the environment, as well as toxicological implications. These reviews can be either general or specific, but properly they may lie in the domains of analytical chemistry and its methodology, biochemistry, human and animal medicine, legislation, pharmacology, physiology, regulation, and toxicology. Certain affairs in food technology concerned specifically with pesticide and other food-additive problems are also appropriate subjects.

Justification for the preparation of any review for this book series is that it deals with some aspect of the many real problems arising from the presence of any foreign chemical in our surroundings. Thus, manuscripts may encompass case studies from any country. Added plant or animal pest-control chemicals or their metabolites that may persist into food and animal feeds are within this scope. Food additives (substances deliberately added to foods for flavor, odor, appearance, and preservation, as well as those inadvertently added during manufacture, packing, distribution, and storage) are also considered suitable review material. Additionally, chemical contamination in any manner of air, water, soil, or plant or animal life is within these objectives and their purview.

Normally, manuscripts are contributed by invitation, but suggested topics are welcome. Preliminary communication with the Editor is recommended before volunteered review manuscripts are submitted.

Tucson, Arizona   G.W.W.

Table of Contents

Rev Environ Contam Toxicol 174:1–18

Ecological Benefits of Contaminated Sediment Remediation

Michael A. Zarull, John H. Hartig, and Gail Krantzberg

Contents

I. Introduction

Sediments contaminated with nutrients, metals, organics, and oxygen-demanding substances can be found in freshwater and marine systems throughout the world. Although some of these contaminants occur in elevated concentrations as a result of natural processes, the presence of many results from human activity. Aquatic sediments with elevated levels of contaminants can be found in any low-energy area that is the recipient of water associated with urban, industrial, or agricultural activity. Such low-energy depositional zones can be found in nearshore embayments and river mouth areas and are also likely to be ecologically significant. These nearshore areas frequently represent the most significant spawning and nursery sites for many species of fish, the nesting and feeding areas for most of the aquatic avian fauna, the areas of highest primary and secondary biological productivity, and the areas of greatest human contact.

Communicated by George W. Ware.

M.A. Zarull (✉)
The United Nations University, International Network on Water, Environment and Health, McMaster University, 1280 Main Street West, Hamilton, Ontario, Canada L85 4L8.

J.H. Hartig
Greater Detroit American Heritage River Initiative, U.S. Coast Guard, Marine Safety Office, 110 Mt. Elliot Ave., Detroit, MI 48207–4380, USA.

G. Krantzberg
International Joint Commission, Great Lakes Regional Office, 100 Ouellette Avenue, Windsor, Ontario N9A 6T3, Canada.

Until recently, determining whether the sediments are causing detrimental ecological impacts and then quantifying the relationships has been limited to indirect or circumstantial evidence (Burton 1992). Although we now have a number of methods available to assess the quality of sediment and its interaction with the rest of the aquatic ecosystem that can be used to estimate ecological risk and even quantify impacts, we cannot accurately measure or predict ecosystem significance on the basis of an examination of the components alone. We still appear to lack an approach that integrates the physical, chemical, and biological components of the ecosystem (Krantzberg et al. 2000).

Sediment removal has been used as a management technique in rivers, lakes, and reservoirs both to reduce the health risks from sediment-associated contaminants and to rehabilitate degraded aquatic ecosystems. The technique has been employed in Asia, Europe, and North America to address nutrient, metal, and persistent organic chemical contamination, with variable success and occasional surprises. In most cases, where some form of sediment remediation has taken place, there has been limited quantification of the sediment–impact linkage before taking action and rather limited monitoring afterward (both temporally and in the ecosystem components examined).

Our purpose is to review the nature and effects of contaminated sediment in aquatic ecosystems, share selected management experiences and the associated ecological response to sediment remediation, and make some recommendations on research and management actions to improve the effectiveness of future remediation projects.

II. Contaminated Sediment in the Aquatic Environment

The accumulation of contaminants in the sediment at levels that are not rapidly lethal may result in long-term, subtle effects to the biota by direct uptake or through the food web. The cycling and bioavailability of sediment-associated contaminants in aquatic systems over both short and long time frames are controlled by physical, chemical, biological, and geological processes.

Physical processes affecting sediment contaminant distribution include mechanical disturbance at the sediment–water interface as a result of bioturbation, advection and diffusion, particle settling, resuspension, and burial. Some examples of significant geological processes affecting contaminant distribution and availability include weathering or mineral degradation, mineralization, leaching, and sedimentation. Chemical processes such as dissolution and precipitation, desorption, and oxidation and reduction can have profound effects, as can biological processes such as decomposition, biochemical transformation, gas production and consumption, cell wall and membrane exchange/permeability, food web transfer, digestion, methylation, and pellet generation. In addition, there are fundamental differences in the physical, chemical, and biological properties and behavior of organic versus inorganic substances (metals, persistent organics, organometals, and nutrients), which suggests the need for a detailed knowledge of the area and the relative importance of these processes before completing an assessment of impact or planning remedial measures to mitigate ecological impairments. Details of the major processes and their effects on contaminant cycling and movement can be found in Forstner and Whittman (1979), Salamons and Forstner (1984), Allan (1986), and Krezovich et al. (1987); however, it is important to explore some of the factors that affect bioavailability and uptake of contaminants, as well as the likely, quantifiable consequences of bioaccumulation.

The rate and mechanism of direct contaminant uptake from sediment by bottom-dwelling organisms can vary considerably among species, and even within species. Factors such as feeding ecology of the organisms, their developmental stage, season, behavior, and history of exposure affect contaminant uptake and body burdens. As well, different routes of uptake (soluble transfers versus contaminated food) can also be expected to affect tissue levels (Russell et al. 1999; Kaag et al. 1998).

Experiments with organochlorine pesticides have yielded conflicting results on the relative significance of diet versus aqueous uptake. Within individual studies, available data on sediment-based bioconcentration factors for various organisms show a wide variation among species for a specific contaminant (Kaag et al. 1998; Roesijadi et al. 1978a,b). Accumulation of both organic and metal contaminants can be passive due to adsorption onto the organism, or it can be an active process driven through respiration. "Case-dwelling" species of benthic invertebrates have been thought less susceptible to contaminants than "free-living" organisms because the bioconcentration factors (BCFs) have been found to be quite different for metals such as copper and zinc. Similar differences have been found for oligochaete and amphipod tissue concentrations for polychlorinated biphenyls (PCBs) and hexachlorobenzene.

Sediment type can profoundly influence the bioavailability of sediment-sorbed chemicals. Many researchers have reported an inverse relationship between chemical availability and sediment organic carbon content (Elder et al. 1996; Augenfield and Anderson 1982). There also appears to be a smaller, not as well defined relationship between sediment particle size and chemical availability. In fine-grained sediment, this most likely due to the increased surface area available for adsorption and the reduced interstitial volume (Adams et al. 1985). Chemicals sorbed to suspensions of organic particles (both living, such as plankton, and nonliving) may constitute sources of exposure for filter-feeding organisms and may be important in deposition. This pathway may be significant, as these organisms have been shown to accelerate the sedimentation processes by efficiently removing and depositing particles contained in the water column (Chen et al. 1999).

Several water quality conditions influence bioaccumulation of contaminants: temperature, pH, redox, water hardness, and physical disturbance. In addition, metals in mixtures may also compete for binding sites on organic molecules, resulting in antagonistic effects (e.g., cadmium and zinc, silver and copper).

The biological community itself can strongly influence the physicochemical environment in the sediment, and in turn, affect the bioavailability of contaminants; for example, primary productivity influences the pH, which can influence metal chemistry; sulfate reduction by bacteria facilitates sulfide formation; the reduction of oxygen by organisms and their activities to anoxia affects redox conditions, and with it, metal redox conversion; organic matter is produced that may complex with contaminants; bioturbation influences sediment–water exchange processes and redox conditions; and methylation of some metals, such as mercury, may occur.

Water-based BCFs indicate that benthic invertebrates generally accumulate to higher concentrations than do fish, which may be attributed to the greater degree of exposure of the benthic invertebrates than fish at the sediment–water interface. Biomagnification occurs when contaminant concentrations increase with successive steps in the trophic structure. However, well-defined trophic levels may not exist in the aquatic ecosystem under examination, especially those experiencing (or that have experienced) anthropogenically generated loadings of various contaminants (Kay 1984; Russell et al. 1999). In addition, individual species may occupy more than one trophic level during the life cycle. These factors not only complicate process and exposure understanding, they also complicate monitoring program designs necessary to document improvement after remediation has taken place. However, "it is no longer sufficient to know only whether chemicals accumulate because bioaccumulation itself is not an effect but a process. Regulatory managers must know whether the accumulation of chemicals is associated with or responsible for adverse affects on the aquatic ecosystem and human health" (USEPA 2000).

It was previously assumed that chemicals sequestered within sediment were unavailable to biota and therefore posed little threat to aquatic ecosystems. Although this is clearly incorrect, the presence of a contaminant (nutrient, metal, or organic) in the sediment does not provide a priori evidence of ecological effect. In addition, a detailed understanding of the relevant processes and a quantification of the associated impacts is critical before developing a management plan.

III. Aquatic Ecosystem Effects of Contaminated Sediment

Although laboratory and field studies are not overwhelming in number, both the risk and the actual impairment to organisms, including humans, have been conclusively established (Geisy and Hoke 1989; Burton 1992; Ingersoll et al. 1997). Biota exposed to contaminated sediment may exhibit increased mortality, reduced growth and fecundity, or morphological anomalies. Studies have also shown that contaminated sediment can be responsible for mutagenic and other genotoxic impairments (Lower et al. 1985; West et al. 1986). These effects are not restricted to benthic organisms—plankton, fish, and humans are also affected both from direct contact and through the food chain.

Metals, in their inorganic forms, do not appear to biomagnify appreciably in aquatic ecosystems; however, methylated forms of metals, such as mercury, do biomagnify. However, the factors controlling the transfer of mercury from the sediment, especially monomethylmercury (the most bioaccumulative form of mercury) to aquatic organisms is poorly understood (Mason and Lawrence 1999). Most persistent toxic organics demonstrate biomagnification to lesser or greater degrees; however, it appears that biomagnification is not as dramatic within aquatic food chains as terrestrial pathways. Also, it appears that where this phenomenon does occur, the biomagnification factors between the lowest and highest trophic levels are usually less than one order of magnitude (Kay 1984). Recent investigations confirm that there is no simple relationship between contaminant concentrations in the sediment and bioavailability; however, observed toxic effects are related to the internal concentrations of certain chemicals (Kaag et al. 1998).

Nuisance algal growth and nutrient relationships in lakes are well documented, with phosphorus being cited as the limiting nutrient in freshwater systems. Some phosphorus is released from the bottom sediment during spring and fall lake circulation in dimictic lakes. In shallow, polymictic lakes, sedimentary phosphorus release may be more frequent, creating greater nuisance problems with the infusion of nutrients to overlying water, especially during summer recreational periods. This influx of nutrients usually results in abundant, undesirable phytoplankton growth, reducing water transparency, increasing color, and in severe cases, seriously depleting dissolved oxygen and potentially leading to fish kills. To prevent this stored release, the bottom sediments need to either be removed (dredged) or isolated from the water column (capped).

Nau-Ritter and Wurster (1983) demonstrated that PCBs desorbed from chlorite and illite particles inhibited photosynthesis and reduced the chlorophyll *a* content of natural phytoplankton assemblages. In a similar study, Powers et al. (1982) found that PCBs desorbed from particles caused reduced algal growth as well as reduced chlorophyll production. The time course for desorption and bioaccumulation appears to be quite rapid, with effects being documented within hours after exposure (Harding and Phillips 1978). The rapid transfer of PCBs and other xenobiotic chemicals from particulate material to phytoplankton has significant ramifications because it provides a mechanism for contaminants to be readily introduced to the base of the food web.

The detrimental effects of contaminated sediment on benthic and pelagic invertebrate organisms have been demonstrated in several laboratory studies. Prater and Anderson (1977a,b), Hoke and Prater (1980), and Malueg et al. (1983) have shown that sediment taken from a variety of lentic and lotic ecosystems was lethal to invertebrates during short-term bioassays. Tagatz et al. (1985) exposed macrobenthic communities to sediment-bound and waterborne chlorinated organics and found similar reductions in diversity to both exposures. Chapman and Fink (1984) measured the lethal and sublethal effects of contaminated whole sediment and sediment elutriates on the life cycle of a marine polychaete and found that both sources were capable of producing abnormalities and mortality; also, reduced-derived benzo[a]pyrene has been shown to result in the formation of potentially mutagenic and carcinogenic metabolites in depositional feeding amphipods (Reichert et al. 1985). Other sublethal effects may be more subtle; for example, infaunal polychaetes, bivalves, and amphipods have been shown to exhibit impaired burrowing behavior when placed in pesticide-contaminated sediment (Gannon and Beeton 1971; Mohlenberg and Kiorboe 1983). Some observations have linked contaminants in sediment with alterations in genetic structure or aberrations in genetic expression. Warwick (1980) observed deformities in chironomid larvae mouthparts that he attributed to contaminants. Wiederholm (1984) showed similar deformities in chironomid mouthparts with occurrences ranging from less than 1% at unpolluted sites (background) to 5%–25% at highly polluted sites in Sweden. Milbrink (1983) has shown setal deformities in oligochaetes exposed to high mercury levels in sediment.

Fish populations may also be impacted by chemicals derived from contaminated sediment. Laboratory studies have shown that fathead minnows held in the presence of contaminated natural sediment may suffer significant mortality (Prater and Anderson 1977 a,b; Hoke and Prater 1980). Morphological anomalies have also been traced to associations of contaminated sediment with fish. Malins et al. (1984) found consistent correlations between the occurrence of hepatic neoplasms in bottom-dwelling fish and concentrations of polynuclear aromatic hydrocarbons (PAHs) in sediment from Puget Sound, Washington. In addition, Harder et al. (1983) have demonstrated that sediment-degraded toxaphene was more toxic than the nondegraded form to the white mullet. These studies illustrate the potential importance of sediment to the health and survival of pelagic and demersal fish species but do not necessarily indicate a cause-and-effect relationship.

We can expect that fish will be exposed to chemicals that desorb from sediment and suspended particles, but the relative contributions of these pathways to any observable biological effects are not obvious. Instead, laboratory bioassays and bioconcentration studies are often required as conclusive supporting evidence. The Elizabeth River, a subestuary of the Chesapeake Bay, is heavily contaminated with a variety of pollutants, particularly PAHs. The frequency and intensity of neoplasms, cataracts, enzyme induction, fin rot, and other lesions observed in fish populations have been correlated with the extent of sediment contamination. In addition, bioaccumulation of these same compounds in fish and resident crabs was also observed. However, essential laboratory studies were not conducted to establish contaminants in sediment as the cause of the observed impairments (USEPA 1998).

There have been few examples of direct impacts of contaminated sediment on wildlife or humans. Some recent studies have established these direct links with ducks and tree swallows (Hoffman et al. 2000; Secord et al. 1999). For the most part, the relationship is largely inferential. Bishop et al. (1995, 1999) found good correlations between a variety of chlorinated hydrocarbons in the sediment and concentrations in bird eggs. They believed this relationship indicated that the female contaminant body burden was obtained locally, just before egg laying. Other studies by Bishop et al. (1991) indicated a link between exposure of snapping turtle (*Chelydra s. serpentina*) eggs to contaminants (including sediment exposure) and developmental success (Bishop et al. 1998). Other investigations of environmentally occurring persistent organics have shown bioaccumulation and a range of effects in the mudpuppy (*Necturus maculosus*) (Bonin et al. 1995; Gendron et al. 1997). In the case of humans (*Homo sapiens*), there is only anecdotal evidence from cases such as Monguagon Creek, a small tributary to the Detroit River, where incidental human contact with the sediment resulted in a skin rash. For the most part, assessments of sediment-associated contaminant impacts on the health of vertebrates (beyond fish) are inferential. This approach is known as risk assessment, and it involves hazard identification, toxicity assessment, exposure assessment, and risk characterization (NAS 1983).

USEPA Superfund risk assessments, which are aimed at evaluating and protecting human health, are designed to evaluate current and potential risks to the "reasonably maximally exposed individual" (USEPA 1989). Both cancer and noncancer health effects for adults and children are evaluated. Data for the evaluation include concentrations of specific chemicals in the sediment, water column, and other media that are relevant to the potential exposure route. These routes of exposure may include ingestion of contaminated water, inhalation of chemicals that volatilize, dermal contact, and fish consumption. The media-specific chemicals of potential concern are characterized on the basis of their potential to cause either cancer or noncancer health effects or both. Once the "hazards" have been identified, the prescribed approach is continued to include toxicity evaluation, exposure assessment, and risk characterization. All of this leads to a potential remedial action, which itself follows a set of prescribed rules.

"Ecological risk assessment (ERA) is the estimation of the likelihood of undesired effects of human actions or natural events and the accompanying risks to nonhuman organisms, populations, and ecosystems" (Suter 1997). The structure of ERA is based on human health risk assessment (HHRA), but it has been modified to accommodate differences between ecological systems and humans. "The principal one is that, unlike HHRA, which begins by identifying the hazard (e.g., the chemical is a carcinogen), ERA begins by dealing with the diversity of entities and responses that may be affected, of interactions and secondary effects that may occur, of scales at which effects may be considered, and of modes of exposure." (Suter 1997). Risk characterization is by weight of evidence. Data from chemical analyses, toxicity tests, biological surveys, and biomarkers are employed to estimate the likelihood that significant effects are occurring or will occur. The assessment requires that the nature, magnitude, and extent of effects on the designated assessment endpoints be depicted. More recent work has focused on the development of, and the relationship between, assessment of measurement endpoints for sediment ecological risk assessments. In addition, scientists active in the field have strongly recommended that a weight-of-evidence approach be used (Ingersoll et al. 1997).

It is apparent that rarely is the relationship between a particular contaminant in the sediment and some observed ecological effect straightforward. Physical, chemical, and biological factors are interactive, antagonistic, and highly dynamic. These factors often preclude a precise quantification of the degree of ecological impairment or effect attributable to a contaminant present in the sediment and, therefore, the degree of ecological improvement or benefit that can be achieved through remediation. Precision in quantifying impairment, remediation, and recovery is always improved through a better understanding of the specifics of ecosystem functioning as well as the behavior of the chemical(s) of concern in that particular ecosystem. "Technically feasible goals require adequate knowledge about basic processes and reliable methods to effect repairs" (Cairns 2000). Although a basic understanding of aquatic ecosystem function and chemical fate is generally available, it is evident that systems appear to be sufficiently unique and our understanding sufficiently lacking that an adaptive management approach to the mitigation of contaminated sediment is the prudent course to follow. This approach requires a much tighter coupling of research, monitoring, and management in every case to develop quantifiable realistic goals and measures of success to achieve these aims.

IV. Ecological Response to Sediment Removal

Although other sediment remediation techniques have been employed, such as capping and *in situ* treatment, sediment removal or dredging has been used longer and more extensively, not only for navigational purposes but also for environmental mitigation. Sediment removal has been used as a management technique in lakes as a means of deepening a lake to improve its recreational potential, to remove toxic substances from the system, to reduce nuisance aquatic macrophyte growth, and to prevent or reduce the internal nutrient cycling that may represent a significant fraction of the total nutrient loading (Larsen et al. 1975). Following are some examples of the removal of sediment contaminated by a nutrient (phosphorus), a metal (mercury), and persistent toxic organic compounds (PCBs and PAHs) from lakes, rivers, and embayments.

A. Nutrients

Lake Trummen, Sweden, is one of the most thoroughly documented dredging projects in the world. An evaluation of the effectiveness of the dredging, whose main purpose was to reduce internal nutrient cycling and enrichment through sediment removal, took place over a time frame of more than 20 yr.

Lake Trummen, with a surface area of approximately 1 km^2, a drainage basin of some 12 km^2, and a mean depth of 2 m, was originally oligotrophic; however, it became hypertrophic after receiving both municipal and industrial discharges over a long period of time. To rectify the problems, both municipal and industrial waste effluents were curtailed in the late 1950s; however, the lake did not recover. In the late 1960s, extensive research was undertaken, resulting in the removal of some 400,000 m^3 of surface sediment (the top meter [1 m], in two 50-cm dredgings) from the main basin in 1970 and 1971.

Bengtsson et al. (1975) indicated that postdredging water column concentrations of phosphorus and nitrogen decreased drastically and that the role of the sediment in recycling nutrients was minimized. Phytoplankton diversity increased substantially, while at the same time their productivity was significantly reduced. The size distribution of phytoplankton also shifted to much smaller cells, and water column transparency more than tripled. The troublesome blue-green algal biomass was drastically reduced, with some nuisance species disappearing altogether (Cronberg 1975). Conditions in the lake had improved to such a degree by the mid-1970s that an additional research and management program was undertaken on the fish community. From the late 1960s throughout the 1980s, an extensive monitoring program was maintained. By the mid-1980s, this program documented a deterioration in water quality and the ecological response. It also helped to ascertain that the changes were from increased atmospheric and drainage basin nutrient inputs.

Similar sediment removal projects have been conducted in other areas: Vajgar pond in the Czech Republic, Lake Herman in South Dakota, and Lake Trehorningen in Sweden, to name just a few. The latter-named project is of particular note, because although there were significant decreases in the water column concentrations of phosphorus, phosphorus concentration remained too high to be algal growth limiting. As a result, algal biomass remained the same as before the dredging was undertaken. This example illustrates the importance of having good understanding and quantification of ecological processes before undertaking a remediation project. In addition, Peterson (1982) noted that through the early 1980s there was little evidence to support the effectiveness of sediment removal as a mechanism of ecological remediation. Lack of supporting research and monitoring data continues to be an obstacle to establishing the effectiveness of sediment cleanups.

B. Metals

Minamata Bay, located in southwestern Japan, is the site of one of the more notorious cases of metal pollution in the environment and its subsequent impacts on human health. A chemical factory released mercury-contaminated effluent into the Bay from 1932 to 1968. In addition to contaminating the water and sediment, methylated mercury accumulated in fish and shellfish. This contamination resulted in toxic central nervous system disease among the individuals who ate these fisheries products over long periods of time. In 1973, the Provisional Standard for Removal of Mercury Contaminated Bottom Sediment was established by the Japanese Environmental Agency. Under this criterion, it was estimated that some 1,500,000 m^3 of sediment would need to be removed from an area of 2,000,000 m^2. Dredging and disposal commenced in 1977 along with an environmental monitoring program to ensure that the activities were not further contaminating the environment. Monitoring included measuring turbidity and other water quality variables, as well as tissue analysis of natural and caged fish for mercury residues. Dredging was completed in 1987, and by 1988 the sampling surveys provided satisfactory evidence that the goals had been achieved. Results of the ongoing monitoring showed that no further deterioration of water quality or increase in fish tissue concentration was occurring. By March of 1990, the confined disposal facility received its final clean cover. The total cost for the project was approximately $40–$42 million U.S.

Postproject monitoring provided clear evidence of a reduction in surficial sediment concentrations of mercury to a maximum of 8.75 mg/kg and an average concentration less than 5 mg/kg (national criterion, 25 mg/kg) (Ishikawa and Ikegaki 1980; Nakayama et al. 1996; Urabe 1993; Hosokawa 1993; Kudo et al. 1998). Mercury levels in fish in the bay rose to their maximum between 1978 and 1981, after the primary source had been cut off and some dredging had begun. Tissue concentrations declined slightly as dredging continued; however, they did fluctuate considerably. Fish tissue levels did finally decline below the target levels of 0.4 mg/kg in 1994, some 4 yr after all dredging activity had ceased (Nakayama et al. 1996). These results demonstrate that mercury in the sediment continued to contaminate the fish and that removal or elimination of that exposure was essential for ecological recovery to occur. It also demonstrates that some impact (increased availability and increased fish tissue concentrations) could be associated with the dredging activity, and that a significant lag time from the cessation of remediation activity was necessary for the target body burdens to be achieved.

C. Persistent Toxic Organic Substances

PCB-Contaminated Sediment Remediation in Waukegan Harbor. Waukegan Harbor is situated in Lake County, Illinois (United States) on the western shore of Lake Michigan. Constructed by filling a natural inlet and portions of adjacent wetlands, Waukegan Harbor has water depths varying from 4.0 to 6.5 m. The harbor sediment is composed of soft organic silt (muck) that lies over medium, dense, fine to coarse sand.

Although substantial recreational use occurs in the area around the harbor, land use in the Waukegan Harbor area is primarily industrial. Of the major facilities present, the Outboard Marine Corporation (OMC) was identified as the primary source of PCB contamination in harbor sediment. U.S. EPA investigations in 1976 revealed high levels of PCBs in Waukegan Harbor sediment and in soil close to OMC outfalls. Concurrently, high levels of PCBs (greater than the U.S. Food and Drug Administration action levels of 2.0 mg/kg PCB) were also found in resident fish species. As a result, in 1981, the U.S. EPA formally recommended that no fish from Waukegan Harbor be consumed.

Remedial efforts in the harbor began in 1990, with harbor dredging conducted in 1992. Approximately 24,500 m^3 of PCB-contaminated sediment was removed from the harbor using a hydraulic dredge. Approximately 2,000 m^3 of PCB-contaminated sediment in excess of 500 mg/kg PCBs was removed from a "hot spot" that accounts for the majority of the PCBs on the site and thermally extracted onsite to at least 97%. Soils in excess of 10,000 mg/kg PCBs were also excavated and treated onsite by thermal extraction (Hartig and Zarull 1991). In all, 11,521,400 kg of material was treated, and 132,500 L PCBs was extracted and taken offsite for destruction, with a total cost of $20–25 million. No soils or sediment that exceeded 50 mg/kg PCBs remained onsite, except those within specially constructed containment cells.

PCB contaminant levels have been monitored in a variety of fish species from Waukegan Harbor on an annual basis since 1978. However, carp have provided the most consistently available information over this period. Fish contaminant monitoring, conducted after the dredging in 1992, showed a substantial decrease for PCB concentrations in carp fillets. Figure 1 presents trend data for PCBs in Waukegan Harbor carp fillets. PCB levels in 1993 fish suggest that dredging did not cause significant PCB resuspension. Contaminant levels in 1993 fish averaged fivefold lower than those tested in previous years through 1991. Contaminant levels from 1993–1995 appeared to remain at these lower levels, but there is a suggestion of an apparent increase for the period 1996–1998. There is no statistically significant difference between the 1983 and 1998 levels of PCBs in carp (based on a two-sample t test).

As a result of the dramatic decline of PCBs in several fish species between the late 1970s and 1990s, the posted Waukegan Harbor fish advisories were removed, although fish advisories still exist for carp and other fish throughout Lake Michigan. The Illinois Lake Michigan Lakewide Advisory is protective of human health, as PCB concentrations in Waukegan Harbor fish are considered similar to those found elsewhere in Lake Michigan.

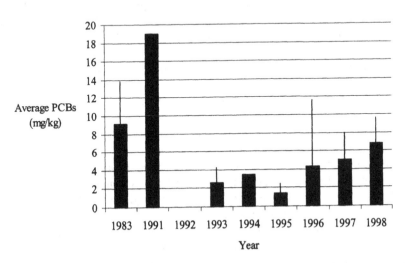

Fig. 1. Average polychlorinated biphenyl (PCB) levels, with 95% confidence intervals, in Waukegan Harbor carp fillets (n = 3–6 each year, except 1991 and 1994, one sample only; 1992 dredging occurred, no sampling). (From U.S. EPA and Illinois EPA, unpublished data.)

PAH-Contaminated Sediment Remediation in the Main Stem, Black River.   The Black River enters the south shore of Lake Erie at Lorain Harbor, in northcentral Ohio. The Black River drainage basin is dominated by agricultural and rural land uses (89%). Residential, commercial, and recreational uses constitute the remaining 11%, concentrated in the lower regions of the river. The area has 45 permitted dischargers, 26 industrial and 19 municipal. The only industrial discharger that is considered to be "major" (discharging >1 million gal/d) by the U.S. EPA is USS/KOBE Steel, located in the lower portion of the river. Until 1982, USS operated a coking facility, which is considered to have been the major source of PAH and metal contamination within the area.

A 1985 Consent Decree (U.S. District Court—Northern District of Ohio 1985) mandated USS/KOBE Steel Company to remove 38,000 m^3 of PAH-contaminated sediment from the main stem of the Black River. The goal of the sediment remediation project was to remove PAH-contaminated sediment to eliminate liver tumors in resident brown bullhead populations.

Tests from 1980 confirmed the presence of elevated levels of cadmium, copper, lead, zinc, cyanide, phenols, PAHs, oils, and grease in sediment adjacent to the former USS steel coke plant outfall. PAH concentrations in this area totaled 1096 mg/kg (Baumann et al. 1982). Tests also confirmed the presence of low levels of pesticides (DDT and its metabolites) in both the main stem and the harbor regions. This sediment exceeded the U.S. EPA Heavily Polluted Classification for Great Lakes harbor sediment. As a result, all main stem and harbor sediment dredged during U.S. Army Corps of Engineers maintenance operations required disposal in a confined disposal facility.

High sediment PAH levels corresponded to a high frequency of liver tumors in resident populations of brown bullheads. Although sediment PAH levels had declined since the USS's coking facility was shut down, levels were still of concern. Sediment remediation occurred upstream of the federal navigational channel in the vicinity of the coke plant outfall. Dredging of the sediment began in 1989. A total of 38,000 m^3 of sediment were removed during the operation. This action was completed in December 1990 and cost approximately $1.5 million for the dredging and containment of the sediment.

The primary cleanup target was the removal of sediment in the area of the former USS coke plant to "hard bottom," or the underlaying shale bedrock. No quantitative environmental targets or endpoints were established, although postdredging sampling was required to test for remaining areas of elevated PAH concentrations. Before dredging, PAH concentrations ranged from 8.8 to 52.0 mg/kg within Black River sediment. As a result of dredging, PAH concentrations in sediment declined (Table 1).

PAH levels in brown bullheads, which had been monitored since the early 1980s (Baumann et al. 1982; Baumann and Harshbarger 1995, 1998), suggest some very interesting relationships between liver neoplasms and the dredging of sediment. Figure 2 illustrates the prevalence of hepatic tissue conditions (cancer, noncancer neoplasm, altered hepatocytes, normal) found in fish of age 3 yr in 1982 (during coke plant operations), 1987 (after coke plant closing, before

Table 1. Polycyclic aromatic hydrocarbon (PAH) concentrations (mg/kg) in Black River sediment in 1980 (during coke plant operations), 1984 (coking facility closed, predredging), and 1992 (postdredging).

PAH compound	1980	1984	1992
Phenanthrene	390.0	52.0	2.6
Fluoranthrene	220.0	33.0	3.7
Benzo[a]anthracene	51.0	11.0	1.6
Benzo[a]pyrene	43.0	8.8	1.7

USS coking facility was closed down in 1982; dredging occurred 1989–1990.

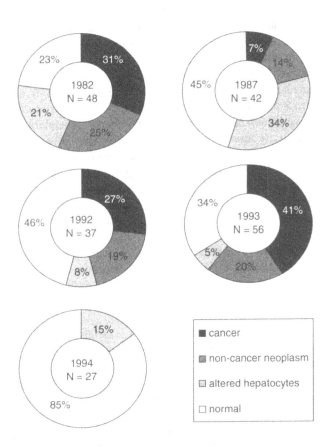

Fig. 2. Percentage of age 3 brown bullheads from the Black River having various liver lesions. 1982 Baseline; 1987 Post Plant Closure (1984); 1992, 1993 and 1994 Post dredging (1990). (Adapted from Baumann and Harshbarger 1998.)

dredging), 1992 (exposed to dredging at age 1), 1993 (exposed to dredging as young of year), and 1994 (hatching after dredging was completed).

The incidence of liver cancer in bullheads of age 3 decreased between 1982 and 1987, corresponding with decreased PAH loadings following the coke plant closure in 1982. There is general consensus that the increase in liver cancer found in the 1992 and 1993 surveys is a result of PAH redistribution, which occurred during the 1990 dredging efforts. No instance of liver cancer was found in 1994 samples of age 3 brown bullheads. Further, the percent of normal liver tissues increased from 34% to 85% between 1993 and 1994. This elimination of liver tumors and the increase in the percentage of normal tissues in the resident brown bullhead populations as a result of sediment remediation provides substantial evidence of the efficacy of the remedial strategy.

Summary

Contaminated sediment has been identified as a source of ecological impacts in marine and freshwater systems throughout the world, and the importance of the contaminated sediment management issue continues to increase in all industrialized countries. In many areas, dredging or removal of sediments contaminated with nutrients, metals, oxygen-demanding substances, and persistent toxic organic chemicals has been employed as a form of environmental remediation. In most situations, however, the documentation of the sediment problem has not been quantitatively coupled to ecological impairments. In addition, the lack of long-term, postactivity research and monitoring for most projects has impeded a better understanding of the ecological significance of sediment contamination.

Establishing quantitatively the ecological significance of sediment-associated contamination in any area is a difficult time- and resource-consuming exercise. It is, however, absolutely essential that it be done. Such documentation will likely be used as the justification for remedial and rehabilitative action(s) and also as the rationale for proposing when intervention is necessary in one place but not another. Bounding the degree of ecological impact (at least semiquantitatively) provides for realistic expectations for improvement if sediment remediation is to be pursued. It should also provide essential information on linkages that could be used in rehabilitating other ecosystem components such as fish or wildlife habitat.

The lack of information coupling contaminated sediment to specific ecological impairments has, in many instances, precluded a clear estimate of how much sediment requires action to be taken, why, and what improvements can be expected to existing impairment(s) over time. Also, it has likely resulted in either a delay in remedial action or abandonment of the option altogether.

A clear understanding of ecological links not only provides adequate justification for a cleanup program but also represents a principal consideration in the adoption of nonintervention, alternative strategies. In developing this understanding, it is important to know not only the existing degree of ecological impairment associated with sediment contaminants but also the circumstances under which those relationships and impacts might change (i.e., contaminants become more available and more detrimental).

Because contaminated sediment remediation often costs millions of dollars per area, adequate assessment, prediction, and monitoring of recovery would seem obvious. However, experience has shown that this is not always the case, particularly for prediction and monitoring of ecological recovery. This scenario would never happen in the business world and should not occur in the environmental management field.

References

Allan RJ (1986) The Role of Particulate Matter in the Fate of Contaminants in Aquatic Ecosystems. IWD Scientific Series No. 142, Burlington, Ontario.

Adams WJ, Kimerle RA, Mosher RG (1985) Aquatic safety assessments of chemicals sorbed to sediments. In: Proceedings of the Seventh Annual Aquatic Toxicology Symposium, Milwaukee, WI, pp 429–453.

Augenfield JM, Anderson JW (1982) The fate of polyaromatic hydrocarbons in an intertidal sediment exposure system: bioavailability to *Macoma inquinata* (Mollusa: Pelecyopoda) and *Abarenicola pacifica* (Annelida: Polychaeta). Mar Environ Res 7: 31–50.

Baumann PC, Harshbarger JC (1995) Decline in liver neoplasms in wild brown bullhead catfish after coking plant closes and environmental PAHs plummet. Environ Health Perspect 103:168–170.

Baumann PC, Harshbarger (1998) Long term trends in liver neoplasm epizootics of brown bullhead in the Black River, Ohio. Environ Monit Assess 53:213–223.

Baumann PC, Smith WD, Ribick M (1982) Hepatic tumor rates and polynuclear aromatic hydrocarbon levels in two populations of brown bullhead (*Ictalurus nebulosus*). In: Cook MW, Dennis AJ, Fisher GL (eds) Polynuclear Aromatic Hydrocarbons, Sixth International Symposium on Physical and Biological Chemistry. Battelle Press, Columbus, OH, pp 93–102.

Bengtsson L, Fleischer G, Lindmark G, Ripl W (1975) Lake Trummen restoration project I. water and sediment chemistry. Verh Int Verein Limnol 19:1080–1087.

Bishop CA, Brooks RJ, Carey JH, Ng P, Norstrom RJ, Lean DRS (1991) The case for a cause-effect linkage between environmental contamination and development in eggs of the common snapping turtle (*Chelydra s. serpentina*) from Ontario, Canada. J Toxicol Environ Health 33:521–547.

Bishop CA, Koster MD, Chek AA, Hussell DJT, Jock K (1995) Chlorinated hydrocarbons and mercury in sediments, red-winged blackbirds (*Agelaius phoneniceus*), and tree swallows (*Tachycineta bicolor*) from wetlands in the Great Lakes-St. Lawrence River Basin. Environ Toxicol Chem 14:491–501.

Bishop CA, Mahony NA, Trudeau S, Pettit KE (1999) Reproductive success and biochemical effects in tree swallows (*Tachycineta bicolor*) exposed to chlorinated hydrocarbon contaminants in wetlands of the Great Lakes and St. Lawrence River Basin, USA and Canada. Environ Toxicol Chem 18:263–271.

Bishop CA, Ng P, Pettit KE, Kennedy SW, Stegeman JJ, Norstrom RJ, Brooks RJ (1998) Environmental contamination and development abnormalities in eggs and hatchlings of the common snapping turtle (*Chelydra serpentina serpentina*) from the Great Lakes–St. Lawrence River Basin (1989–91). Environ Pollut 101:143–156.

Bonin J, DesGranges JL, Bishop CA, Rodrigue J, Gendron A (1995) Comparative study of contaminants in the mudpuppy (Amphibia) and the common snapping turtle (Reptilia), St. Lawrence River, Canada. Arch Environ Contam Toxicol 28:184–194.

Burton GA Jr (ed) (1992) Sediment Toxicity Assessment. Lewis, Chelsea, MI.

Cairns J Jr (2000) Setting ecological restoration goals for technical feasibility and scientific validity. Ecol Eng 15:171–180.

Chapman PM, Fink R (1984) Effects of Puget Sound sediments and their elutriates on the life cycle of *Capitella capita*. Bull Environ Contam Toxicol 33:451–459.

Chen W, Kan AT, Fu G, Vignona LC, Tomson MB (1999) Adsorption-desorption behaviors of hydrophobic organic compounds in sediments of Lake Charles, Louisiana, USA. Environ Toxicol Chem 18:1610–1616.

Cronberg G, Gelin C, Larsson K (1975) Lake Trummen restoration project II. Bacteria, phytoplankton and phytoplankton productivity. Verh Int Verein Limnol 19:1088–1096.

Elder JF, James RV, Steuer JJ (1996) Mobility of 2,2′,5,5′-tetrachlorobiphenyl in model systems containing bottom sediments and water from the lower Fox River, Wisconsin. J Great Lakes Res 22:697–706.

Forstner U, Whittman GTW (1979) Metal Pollution in the Aquatic Environment. Springer, Berlin.

Gannon JE, Beeton AM (1971) Procedures for determining the effects of dredged sediments on biota—benthos viability and sediment selectivity tests. J Water Pollut Control Fed 43:392–398.

Gendron AD, Bishop CA, Fortin R, Hontela A (1997) *In vitro* testing of the functional integrity of the corticosterone-producing axis in mudpuppy (Amphibia) exposed to chlorinated hydrocarbons in the wild. Environ Toxicol Chem 16:1694–1706.

Giesy JP, Hoke RA (1989) Freshwater sediment toxicity bioassessment: rationale for species selection and test design. J Great Lakes Res 15:539–569.

Harder HW, Carter TV, Bidleman TF (1983) Acute effects of toxaphene and its sediment-degraded products on estuarine fish. Can J Fish Aquat Sci 40:2119–2125.

Harding LW, Phillips JH (1978) Polychlorinated biphenyls: transfer from microparticulates to marine phytoplankton and the effects on photosynthesis. Science 202:1189–1192.

Hartig JH, Zarull MA (1991) Methods of restoring degraded areas in the Great Lakes. Rev Environ Contam Toxicol 117:127–154.

Hoffman DJ, Heinz GH, Sileo L, Audet DJ, Campbell JK, LeCaptain LJ (2000) Developmental toxicity of lead-contaminated sediment to mallard ducklings. Arch Environ Contam Toxicol 39:221–232.

Hoke RA, Prater BL (1980) Relationship of percent mortality of four species of aquatic biota from 96-hour sediment bioassays of five Lake Michigan harbors and elutriate chemistry of the sediments. Bull Environ Contam Toxicol 25:394–399.

Hosokawa Y (1993) Remediation work for mercury contaminated bay: experiences of Bay Project, Japan. Water Sci Technol 28:339–348.

Ingersoll CG, Dillon T, Biddinger GR (eds) (1997) Ecological Risk Assessment of Contaminated Sediments. SETAC Pellston Workshop on Sediment Ecological Risk Assessment. SETAC Press, Pensacola, FL.

International Joint Commission (IJC) (1999) Protecting what has been gained in the Black River. A report on a public symposium held in Lorain, Ohio, October 8, 1998. IJC, Windsor, Ontario, Canada.

Ishikawa T, Ikegaki Y (1980) Control of mercury pollution in Japan and the Minamata Bay cleanup. J Water Pollut Control Fed 52:1013–1018.

Kaag NHBM, Foekema EM, Scholten MCT (1998) Ecotoxicity of contaminated sediments, a matter of bioavailability. Water Sci Technol 37:225–231.

Kay SH (1984) Potential for biomagnification of contaminants within marine and freshwater food webs. Technical Report D-84-7. Department of the Army, Waterways Experiment Station, Corps of Engineers, Vicksburg, MS.

Krantzberg G, Reynoldson T, Jaagumagi R, Bedard D, Painter S, Boyd D, Pawson T (2000) SEDS: setting environmental decisions for sediment, a decision making tool for sediment management. Aquat Ecosyst Health Manage 3:387–396.

Krezovich JP, Harrison FL, Wilhelm RG (1987) The bioavailability of sediment-sorbed organic chemicals: a review. Water Air Soil Pollut 32:233–245.

Kudo A, Fujikawa Y, Miyahara S, Zheng J, Takigami H, Sugahara M, Muramatsu T (1998) Lessons from Minamata mercury pollution, Japan: after a continuous 22 years of observation. Water Sci Technol 38:187–193.

Larsen DP, Malueg KW, Schults D, Brice RM (1975) Response of eutrophic Shagawa Lake, Minnesota, U.S.A., to point-source, phosphorus reduction. Verh Int Verein Limnol 19:884–892.

Lower WR, Yanders AF, Marrero TR, Underbring AG, Drobner VK, Collins MD (1985) Mutagenicity of bottom sediment From a water reservoir. Environ Toxicol Chem 4: 13–19.

Malins DC, McCain BB, Brown DW, Varanasi U, Krahn MM, Myers MS, Chan SL (1984) Sediment-associated contaminants and liver diseases in bottom-dwelling fish. Hydrobiologia 149:67–74.

Malueg KW, Schuytema GS, Gakstatter JH, Krawczyk DF (1983) Effect of *Hexagenia* on *Daphnia* responses in sediment toxicity tests. Environ Toxicol Chem 2:73–82.

Mason RP, Lawrence AL (1999) Concentration, distribution, and bioavailability of mercury and methylmercury in sediments of Baltimore Harbor and Chesapeake Bay, Maryland, USA. Environ Toxicol Chem 18:2438–2447.

Milbrink G (1983) Characteristic deformities in tubificid oligochaetes inhabiting polluted bays of Lake Vanern, southern Sweden. Hydrobiologia 106:169–184.

Mohlenberg F, Kiorboe T (1983) Burrowing and avoidance behaviour in marine organisms exposed to pesticide-contaminated sediment. Mar Pollut Bull 14:57–60.

Nakayama Y, Nakai O, Nanba T, Kyuumak K (1996) Effect of the Minamata Bay environment restoration project. Proceedings, 17th US/Japan Experts Meeting: Management of Bottom Sediments Containing Toxic Substances, 12–14 March, 1996, Oakland, CA.

National Academy of Sciences (NAS) (1983) Risk Assessment in the Federal Government: Managing the Process. National Academy Press, Washington, DC.

Nau-Ritter GM, Wurster CF (1983) Sorption of polychlorinated biphenyls (PCBs) to clay particulates and effects of desorption on phytoplankton. Water Res 17:383–387.

Peterson SA (1982) Lake restoration by sediment removal. Water Res Bull 18:423–435.

Powers CD, Nau-Ritter GM, Rowland RG, Wurster CF (1982) Field and laboratory studies of the toxicity to phytoplankton of polychlorinated biphenyls (PCBs) desorbed from fine clays and natural suspended particulates. J Great Lakes Res 8:350–357.

Prater BL, Anderson MA (1977a) A 96-hour bioassay of Otter Creek, Ohio. J Water Pollut Control Fed 49:2099–22106.

Prater BL, Anderson MA (1977b) A 96-hour sediment bioassay of Duluth and Superior Harbor basins (Minnesota) using *Hexagenia lamboidea, Assails communis, Daphnia magna,* and *Pimephales promelas* as test organisms. Bull Environ Contam Toxicol 18:159–169.

Reichert WL, Le Eberhart BT, Varanasi U (1985) Exposure of two species of deposit-feeding amphipods to sediment-associated [^{3}H] benzo[a]pyrene: uptake, metabolism, and covalent binding to tissue macromolecules. Aquat Toxicol 3:45–56.

Roesijadi G, Anderson JW, Blaylock JW (1978a) Uptake of hydrocarbons from marine sediments contaminated with Prudhoe Bay crude oil: influence of feeding type of test species and availability of polycyclic aromatic hydrocarbons. J Fish Res Board Can 35:608–614.

Roesijadi G, Woodruff DL, Anderson JW (1978b) Bioavailability of napthalenes from marine sediments artificially contaminated with Prudhoe Bay crude oil. Environ Pollut 15:223–229.

Russell RW, Gobas RAPC, Haffner GD (1999) Role of chemical and ecological factors in trophic transfer of organic chemicals in aquatic food web. Environ Toxicol Chem 18:1250–1257.

Salamons W, Forstner U (1984) Metals in the Hydrocycle. Springer, Berlin.

Secord AL, McCarty JP, Echols KR, Meadows JC, Gale RW, Tillitt DE (1999) Polychlorinated biphenyls and 2,3,7,8-tetrachlorodibenzo-*p*-dioxin equivalents in tree swallows from the upper Hudson River, New York State, USA. Environ Toxicol Chem 18:2519–2525.

Suter GW II (1997) Overview of the ecological risk assessment framework. In: Ingersoll CG, Dillon T, Biddinger GR (eds) Ecological Risk Assessments of Contaminated Sediments. SETAC Press, Pensacola, FL, pp 1–6.

Tagatz ME, Plaia GR, Deans CH (1985) Effects of 1,2,4-trichlorobenzene on estuarine macrobenthic communities exposed via water and sediment. Ecotoxicol Environ Saf 10:351–360.

United States Environmental Protection Agency (USEPA) (1998) EPA's contaminated sediment management strategy. EPA-823-R-98–001. USEPA Office of Water, Washington, DC.

United States Environmental Protection Agency (USEPA) (1989) Risk assessment guidance for Superfund. Volume 1. Human health evaluation manual. Part A. Interim final. EPA/540/1–89/002. USEPA, Office of Emergency and Remedial Response, Washington, DC.

United States Environmental Protection Agency (USEPA) (2000) Bioaccumulation testing and interpretation for the purpose of sediment quality assessment: status and needs. EPA-823-R-00–001. USEPA Office of Water, Washington, DC.

Urabe S (1993) Outline of mercury sediment work in Minamata Bay. Proceedings of the 16th U.S./Japan Experts Meeting: Management of Bottom Sediments Containing Toxic Substances, 12–14 October, 1993, Kitukyushu, Japan.

Warwick WF (1980) Paleolimnology of the Bay of Quinte, Lake Ontario: 2800 years of cultural influence. Can Bull Fish Aquat Sci 206:1–117.

West WR, Smith PA, Booth GM, Lee ML (1986) Determination and genotoxicity of nitrogen heterocycles in a sediment from the Black River. Environ Toxicol Chem 5:511–519.

Wiederholm T (1984) Incidence of deformed chironomid larvae (Diptera: Chironomidae) in Swedish lakes. Hydrobiologia 109:243–249.

Manuscript received May 16; accepted May 18, 2001.

Rev Environ Contam Toxicol 174:19–48

Environmental Fate of Triclopyr

Allan J. Cessna, Raj Grover†, and Don T. Waite

Contents

I. Introduction

Triclopyr, a postemergence herbicide, was first reported in 1975 by Byrd and coworkers. It is a selective systemic herbicide that is rapidly absorbed by the foliage and roots of plants; for example, uptake into wheat and barley leaves was

---

Communicated by George W. Ware.

A.J. Cessna
Research Centre, Agriculture and Agri-Food Canada, P.O. Box 3000, Lethbridge, Alberta T1J 4B1 Canada.
National Water Research Institute, 11 Innovation Boulevard, Saskatoon, Saskatchewan S7N 3H5 Canada.

R. Grover
(Deceased) Formerly Director, Research Station, Agriculture and Agri-Food Canada, Regina, Saskatchewan, Canada.

D.T. Waite
Environment Canada, 300-2365 Albert Street, Regina, Saskatchewan S4P 4K1 Canada

essentially complete 12 hr after treatment (Lewer and Owen 1990). Triclopyr is rapidly translocated throughout plants (Gorrell et al. 1988; Lewer and Owen 1990), primarily by the symplastic pathway, and accumulates in meristematic tissue (Radosevich and Bayer 1979). The herbicide induces auxin-type responses in susceptible plant species that include epinastic bending and twisting of stems and petioles, swelling and elongation of stems, and cupping and curling of leaves. These effects are followed by chlorosis at the growing points, growth inhibition, wilting, and necrosis. Death of susceptible plants occurs slowly, usually within 3–5 wk.

At rates of 1–10 kg acid equivalent (a.e.) ha^{-1}, triclopyr can be applied to control woody plants and many broad-leaved weeds in noncropland areas such as rights-of-way, industrial areas, coniferous forests, rangeland, and permanent pastures. In forestry, it is also used for site preparation and conifer release. At lower application rates (0.13–1.0 kg a.e. ha^{-1}), it is used for weed control in rice, turf, and plantation crops (palm oil, rubber).

II. Identity and Physicochemical Characteristics
A. Identity

The IUPAC chemical name for triclopyr (Fig. 1) is 3,5,6-trichloro-2-pyridyloxy-acetic acid, whereas the Chemical Abstracts name is [(3,5,6-trichloro-2-pyridi-nyl)oxy]acetic acid. It is known by the trade name Garlon® and was introduced by Dow Chemical Co. by its code name 'Dowco 233.' Other trade names are Turflon® and Grazon®.

Triclopyr is a substituted acetic acid (Fig. 1) that has the empirical formula $C_7H_4Cl_3NO_3$ and a molecular weight of 256.5 g mol^{-1}. It is a fluffy colorless solid with a melting point range from 148° to 150 °C and is stable under normal storage conditions (Anonymous 1994). Its CAS registration no. is 55335–06-3. It has a pK_a value (2.93; Wolt 1998) similar to those of other aryloxyacetic acids (Cessna and Grover 1978) and is generally formulated either as the water-soluble triethylamine (triclopyr TEA) or isopropylamine (triclopyr IPA) salt, or the oil-soluble 2-butoxyethyl ester (triclopyr BEE).

B. Physicochemical Characteristics

The chemical and physical properties of triclopyr, triclopyr TEA, and triclopyr BEE are listed in various handbooks (Anonymous 1994, 1997). Wauchope et al. (1992) and Wolt (1998) also list their environmental parameters. Environ-

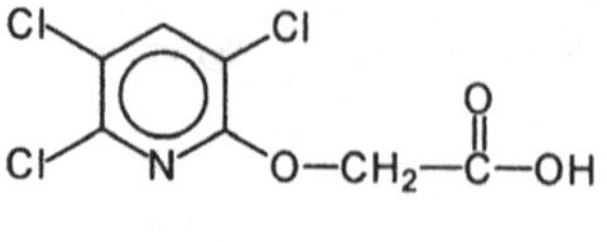

Fig. 1. Structural formula for triclopyr.

Table 1. Water solubilities, vapor pressures, Henry' law constants, field half-lives, and adsorption coefficients of triclopyr, triclopyr 2-butoxyethyl ester (BEE), and triclopyr triethylamine (TEA).[a]

Triclopyr	Water solubility, S_w (mg L^{-1})	Vapor pressure, P_{vap} (mPa)	Henry's law constant, K_H (atm m^3 mol^{-1})	Field half-life, $t_{1/2}$ (d)	Adsorption coefficient, K_{oc} (L kg^{-1})
Acid[b]	430	0.17	9.65×10^{-10}	32	20[c]
BEE	6.8	0.48	2.47×10^{-7}	1.1[d]	780[c]
TEA	4.12×10^5	<0.13	1.15×10^{-14}	—[e]	—

[a]All values are from Wolt (1998) unless otherwise indicated; [b]Exists in anionic form at pH values typical of natural waters (pH 5–9); [c]Value is from Anonymous (1997); [d]The 2-butoxyethyl ester is rapidly hydrolyzed to the acid; [e]The TEA salt of triclopyr dissociates extremely rapidly to the acid.

mentally relevant physical parameters (water solubility, vapor pressure, Henry's law constant, field half-life, and soil adsorption coefficient) of triclopyr and its 2-butoxyethyl ester and triethylamine salt are summarized in Table 1.

III. Transformation Processes
A. Hydrolysis

Triclopyr is stable to abiotic hydrolysis. Negligible degradation of triclopyr occurred during a period of 30 d at pH 5, 7, and 9 (Cleveland and Holbrook 1991) or for periods greater than 200 d at 35 °C at pH 4.5–8.5 (Szeto 1993).

Triclopyr BEE is susceptible to abiotic hydrolysis. Bidlack (1978), using quantitative thin-layer chromatography (TLC) techniques, reported that the rate of hydrolysis of triclopyr BEE in phosphate-buffered water increased with increasing pH and temperature (Table 2; data reported in McCall et al. 1988), and

Table 2. Hydrolysis of triclopyr BEE in phosphate-buffered water as a function of pH and temperature.[a]

pH	Temperature (°C)	Half-life (d)
5	15	208
	25	84
7	15	25.5
	25	8.7
9	15	1.7
	25	0.3
6.7[b]	25	0.5

[a]Data reported in McCall et al. (1988).
[b]Natural unbuffered creek water.

Table 3. Hydrolysis of triclopyr BEE in three soils under aerobic conditions.

Time (hr)	Percent triclopyr BEE remaining			
	Catlin silty clay loam (pH 5.4)	Commerce silty loam (pH 6.1)	Londo sandy loam (pH 6.8)	Mean
0	100	100	100	100
4	28	25	25.5	26.2
6	18.2	16.4	15.1	16.6
16	12.4	8.4	9.2	10
48	3.2	2.8	1.8	2.7
Half-life (hr)[a]	2.4	2.3	2.2	2.3

[a]Half-lives calculated during the first 6 hr.

that ester hydrolysis was rapid (at 25 °C) in moist soil and natural water, with half-lives of 3 and 0.5 hr, respectively. Consequently, it was concluded that only triclopyr need be considered for environmental impact studies. Since then, laboratory (McCall et al. 1988; Szeto 1993) and field (Petty and Gardner 1993; Thompson et al. 1995) studies have confirmed this observation, both in natural waters and in moist soils.

Laboratory Studies. Using three soils in a model aquatic system, representing a shallow, static pond, McCall et al. (1988) reported that the hydrolysis half-lives of triclopyr BEE in sediments under aerobic conditions were ~2 hr (Table 3). Under anaerobic conditions, the hydrolysis was also as rapid, indicating that the ester was inherently unstable in moist soils or sediments. Szeto (1993), using high pressure liquid chromatography (HPLC) methods, also demonstrated the dependence of the hydrolysis rate of the triclopyr ester on pH in phosphate- and borate-buffered water (Table 4) and the rapid hydrolysis of the ester in river waters.

Field Studies. In a study in which triclopyr BEE was injected into a small forest stream, the half-life of the ester was ~1 hr (Thompson et al. 1995). Following field application to a sandy loam soil, triclopyr BEE hydrolyzed rapidly to triclopyr acid with a half-life of approximately 1.1 d (Petty and Gardner 1993).

B. Photodecomposition

For a photoreaction to occur in the environment, a compound must absorb solar light energy. This may involve direct absorption of a photon by the compound or a sensitized reaction in which reaction of the compound is initiated through light absorption by a chromophore other than the compound itself. Because of

Table 4. Hydrolysis of triclopyr BEE in buffered solution at several pH values and in natural water at 35 °C.

Medium	pH	Half-life (hr)
Phosphate-buffered water	4.5	1909
Phosphate-buffered water	5.5	1313
Phosphate-buffered water	6.5	200
Phosphate-buffered water	7.5	24.5
Borate-buffered water	8.5	4.3
Capilano River	6.6	281
Tamihi Creek	7.6	40.1

a thin layer of ozone in the upper atmosphere, which absorbs almost all of the sun's emitted radiation below 290 nm, it is generally accepted that the only light from the sun available for absorption is that at wavelengths longer than 290 nm. Because triclopyr absorbs light at wavelengths over the range 290–320 nm (Mc-Call and Gavit 1986), it is susceptible to sunlight photolysis.

The artificial sunlight photolysis half-lives of triclopyr and triclopyr BEE in pH 5 phosphate buffer solutions were 5.4 and 33.1 d, respectively (McCall and Gavit 1986). Calculated half-lives of triclopyr and its BEE ester in natural waters at 40 ° N latitude ranged from 2.8 to 14.1 hr and from 16.7 to 83.4 hr (Table 5), respectively, depending on the season of the year and the depth of water. Photodegradation of triclopyr was approximately six times faster than that of the ester, with calculated midday, midsummer half-lives of 2.1 and 12.5 hr, respectively, at the water surface. Calculated half-lives at 1-m depth were somewhat longer ($\sim$30%), irrespective of the season.

Woodburn et al. (1993a) reported that pseudo-first-order half-lives of triclo-

Table 5. Calculated midday, seasonal half-lives of triclopyr and its butoxyethyl ester at pH 5.2, 35 °C, and 40 ° N latitude.

Season	Water depth	Half-life (hr)	
		Triclopyr	Triclopyr BEE
Spring	Surface	2.8	16.7
	1 m	3.7	22.0
Summer	Surface	2.1	12.5
	1 m	2.8	16.5
Fall	Surface	4.6	27.6
	1 m	6.2	36.7
Winter	Surface	10.6	63.5
	1 m	14.1	83.4

pyr in pH 7 buffered water, when exposed to either sunlight or artificial sunlight, were 0.6 and 0.36 d, respectively (Table 6). Previous studies of the photolysis of aqueous solutions of chlorinated pyridine compounds such as picloram (Skurlatov et al. 1983; Woodburn et al. 1989) have indicated that low molecular weight carboxylic acids such as oxamic acid may be produced. In the case of triclopyr, a single major photoproduct was identified as 5-chloro-3,6-dihydroxy-2-pyridinyloxyacetic acid (MDPA), along with minor amounts of oxamic acid and other low molecular weight carboxylic acids (Woodburn et al. 1993a). Photolysis with artificial sunlight resulted in a distribution of photoproducts similar to that obtained with sunlight photolysis.

However, when triclopyr was photolyzed in river water, MDPA was present only as a minor product. In contrast, oxamic acid was the major photoproduct, with several low molecular weight organic acids as additional minor products. The low molecular weight acids were tentatively identified as oxalic acid, pyruvic or maleic acid, and malic or 2-chlorosuccinic acid. Differences in photoproduct distributions in the river water and the pH 7 buffered water were attributed to the presence of dissolved organic material such as humic and fulvic acids in the river water. Somewhat longer half-lives (0.71 and 1.86 d for triclopyr and triclopyr BEE, respectively) were observed in the natural river water.

Trichloropyridinol (TCP), the major degradation product of triclopyr in soil under aerobic conditions (see biological transformation section), readily underwent artificial sunlight photolysis in aqueous phosphate buffer solution (Dilling et al. 1984). The SOLAR simulation model of Zepp and Cline (1977) predicted that, in midsummer sunlight, the half-life of TCP in river water at 1-m depth and 40° N latitude would be approximately 2 hr (Dilling et al. 1984). As demonstrated by this prediction, TCP should undergo rapid photolysis in water.

C. Degradation in Soil

Laboratory Studies.   Bidlack et al. (1977) incubated ring-labeled ^{14}C-triclopyr in two agricultural soils [Commerce loam (0.86 % organic matter, pH 6.6) and Flanagan silty clay loam (2.1% organic matter, pH 5.2)] at 25 °C under

Table 6. First-order photodegradation kinetics of triclopyr in sterile, pH 7 buffered water, and in natural river water, maintained at 25 °C.

Water	Light	Half-life (d)
Sterile, pH 7[a]	Artificial[b]	0.36
	Sunlight[c]	0.60
River water	Artificial	0.71
	Sunlight	1.86

[a]Phosphate buffer solution.
[b]Irradiance spectrum from 290 to 400 nm.
[c]Midsummer sun (40° N latitude).

aerobic and anaerobic (waterlogged) conditions, and then isolated the resulting degradation products using TLC techniques. In both soils, the major intermediate degradation product was 3,5,6-trichloro-2-pyridinol (TCP) (Fig. 2). A second degradation product, present in small amounts, was 3,5,6-trichloro-2-methoxy-pyridine (TMP). Radiolabeled carbon dioxide, which indicated degradation of the pyridine ring, was the only other degradation product identified. Highly polar degradation products, which remained at the origin of the TLC plates, generally accounted for approximately 1% or less of the radioactivity applied as triclopyr. Under aerobic conditions, the half-lives for the degradation of triclopyr in the Commerce and Flanagan soils were 18 and 8 d, respectively. Under anaerobic (waterlogged) conditions, half-lives of triclopyr were longer, with the corresponding values for the two soils being 130 and 42 d.

The degradation of TCP was also much more rapid in the Commerce loam and Flanagan silty clay loam soils under aerobic conditions (Bidlack 1977), which, at least in part, accounted for the much greater evolution of ^{14}C-CO_2 during the degradation of triclopyr under aerobic conditions. Under waterlogged conditions, the much slower degradation of TCP resulted in its accumulation in the soil and in the layer of water over the soil. Bidlack (1977) studied the degradation of TCP in 15 soils that represented 10 major agricultural soil groups in the United States. In all cases, following incubation of ring-labeled ^{14}C-TCP in the soils, the major degradation product was CO_2 with the degradation product TMP being present in amounts varying from 15% to 24%. The ability of fungi, such as *Aspergillus* and *Penicillium* species, to methylate phenols

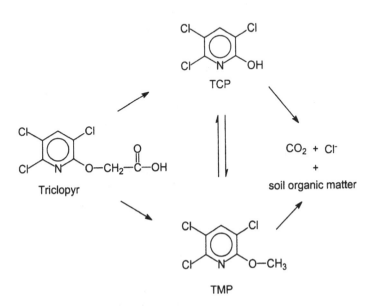

Fig. 2. Proposed pathway for the degradation of triclopyr in soil.

(Cserjesi and Johnson 1972; Curtis et al. 1972) may account for the formation of TMP.

Soil incubation studies have indicated that there was no appreciable degradation of triclopyr under aerobic sterile soil conditions (McCall et al. 1976) compared to that in nonsterile soil (McCall et al. 1976; Bidlack et al. 1977), suggesting the role of microbial activity in its degradation. More recently, Feng (1995) has isolated *Pseudomonas* sp., which are known to readily utilize TCP as a sole source of carbon, from triclopyr-treated soil.

Bidlack et al. (1977) proposed a tentative pathway for the degradation of triclopyr in soil (see Fig. 2) in which TCP was the primary degradation product. Some TCP may be directly methylated to TMP by soil fungi or, alternatively, a small portion of triclopyr may be degraded to TMP by decarboxylation of the acetate moiety. TCP and TMP are then further degraded.

Field Studies.   Poletika and Phillips (1996) studied the dissipation of triclopyr in soil and water following two sequential applications to rice and the subsequent rise and decline of the two major aerobic soil metabolites TCP and TMP. The first herbicide application was before flooding and the second after flooding. Concentrations of TCP and TMP in the water were typically two to three orders of magnitude less than that of triclopyr. First-order half-lives of triclopyr for the preflood and postflood applications varied from 2.2 to 7.6 d and from 1.8 to 3.4 d, respectively. The postflood half-lives are similar to that (4 d) reported by Johnson et al. (1995). Corresponding values for TCP were 1.0–1.2 d and 0.3–0.7 d, whereas those for TMP were 0.4–0.5 d and 2.7–4.1 d, respectively. The rapid dissipation of both triclopyr and TCP suggests that photodegradation may have been the dominant dissipation process. When the water was drained from the paddies, concentrations of triclopyr were 15 ng L^{-1} or less, TCP concentrations were <1 ng L^{-1}, and TMP was not detected. Dissipation of triclopyr and TCP was slower in soil. First-order half-lives of triclopyr during flooding ranged from 2.9 to 12 d and from 117 to 158 d during the postdrainage period. Corresponding values for TCP were 35–50 d and 89–266 d.

When triclopyr BEE was applied to a sandy loam soil, the half-life of triclopyr acid (following the very rapid hydrolysis of the ester; half-life, 1.1 d) was calculated to be 10.6 d (Petty and Gardner 1993), which was somewhat greater than that (5 d) reported by Johnson et al. (1995) when triclopyr was applied to silty loam and silty clay soils. When applied to a cut field of red clover, timothy, and fescue grown in sandy loam soil, triclopyr had a half-life of 15 d (Wolt et al. 1991). In both studies, TCP and, to a lesser extent, TMP were detected as degradation products of triclopyr.

D. Metabolism in Plants

Laboratory Studies: The metabolism of triclopyr in plants has been investigated in only a few studies. In a greenhouse study, Bovey et al. (1979) foliarly treated huisache plants with triclopyr TEA and noted that concentrations of triclopyr in the leaves and stems generally decreased with time. Lewer and Owen (1989) studied the uptake and metabolism of ^{14}C-triclopyr in soybean cell suspension cultures. At 7 d after treatment, two major metabolites resulted from amino acid conjugation, and these metabolites were identified as the aspartate and glutamate amide conjugates. Following treatment of wheat, barley, and chickweed seedlings with ^{14}C-triclopyr BEE, Lewer and Owen (1990) observed extremely rapid hydrolysis of the ester by all three species. Metabolism of the resulting triclopyr was extensive, ranging from 75% to 90% after 7 d. The half-life of triclopyr in wheat was <12 hr, ~24 hr in barley, and >48 hr in chickweed. The dominant metabolite from chickweed, identified as triclopyr aspartate, may have also been formed in wheat. In wheat and barley, triclopyr was metabolized predominantly to complex mixtures of compounds having the properties of sugar esters, but none were specifically identified.

Field Studies.   The dissipation of triclopyr in plants has been investigated in several studies that provide indirect evidence for triclopyr metabolism in plants. In an early study, Radosevich and Bayer (1979) made foliar applications of ^{14}C-triclopyr to bigleaf maple, tanoak, snowbush ceanothus, and bean and barley seedlings and noted that, at 4 d after treatment, recovery of triclopyr ranged from 63% to 87% from these five species. Whisenant and McArthur (1989) treated several woody species in an Idaho forest with triclopyr BEE and reported that triclopyr residues in terminal branch and leaf segments of this vegetation continuously decreased with time. Newton et al. (1990) reported that triclopyr dissipated more slowly in tanoak when applied as the BEE (half-life, 73.5–127.3 d) compared to the TEA salt (18.9–29.0 d). When triclopyr BEE was applied to pasture grass, the dissipation half-life for triclopyr was ~30 d over a 249-d period (Wilcock et al. 1991). A similar dissipation half-life (~40 d over a 365-d period) in pasture grass was reported by Norris et al. (1987), who observed a rapid initial dissipation over the first 2 wk and a much slower rate of loss over the remainder of the year. Whisenant and McArthur (1989) also noted a very rapid initial dissipation of triclopyr in forest grasses (dissipation half-life, ~3 d). The dissipation of triclopyr in plants reported in these studies most likely reflects not only metabolism but also rainfall washoff, volatility losses, and photodegradation.

In contrast, Gorrell et al. (1988) applied ^{14}C-triclopyr mixed with a commercial formulation of triclopyr TEA to horsenettle and observed little or no metabolism of the herbicide. Following aerial application of triclopyr BEE to a forest stream, Thompson et al. (1991) reported that the dissipation half-life of triclopyr in two aquatic species (sedge and broad-leaved arrowhead) was 0.4–0.7 d. Injection of triclopyr TEA into Lake Seminole resulted in a dissipation half-life of ~4 d in two other aquatic plants (Eurasian watermilfoil and hydrilla) (Woodburn et al. 1993b). Thus, the rate of dissipation of triclopyr in plants is species dependent.

Triclopyr metabolites in plants have been identified in only two field studies. Following application of triclopyr IPA to pasture grass, Norris et al. (1987)

detected <1% conversion of triclopyr to TCP and TMP. TCP was detected continuously in the grass tissue over a 365-d period, whereas TMP was detected
only during the first week of the study. Trace amounts (< 0.05 mg kg^{-1}) of TCP
were detected in aquatic plants (Eurasian watermilfoil and hydrilla) for 3 d
following application of triclopyr TEA to Lake Seminole (Woodburn et al.
1993b).

IV. Mobility

A. Adsorption/Desorption

A number of soil sorption studies have established the relatively greater adsorptive behavior of triclopyr BEE compared to that of triclopyr and its metabolite
TCP. Because of the rapid hydrolysis of triclopyr BEE in moist soil (Petty and
Gardner 1993), McCall et al. (1988) carried out 5-min batch adsorption studies
in four soils (Table 7). Triclopyr BEE was shown to be strongly adsorbed by
all four soils. Soil adsorption values (K_d) ranged from 10 to 68 L kg^{-1}, and the
average value for K_{oc} was approximately 1200 L kg^{-1}.

In contrast, triclopyr and its metabolite TCP were reported to be weakly
adsorbed to soils similar to those studied by McCall et al. (1988). The average
K_d value for triclopyr in 4 soils was 0.61 L kg^{-1} and the corresponding K_{oc} value
was 59 L kg^{-1} (Woodburn et al. 1988) (see Table 7). Racke (1993) reported K_d
and K_{oc} values for TCP in 26 soils. The average K_d value for these soils was 2.9
L kg^{-1} whereas that for K_{oc} was 159 L kg^{-1}. The soil sorption characteristics of
TMP, the soil metabolite that is formed in lesser amounts than TCP, were similar to those of triclopyr BEE (Racke 1993).

By determining environmentally relevant K_d values (corresponding to worst
case scenarios) for triclopyr in 4 soils and using average application rates under
field conditions, Wolt (1998) showed that adsorption of triclopyr in soil was
strongly related to soil organic carbon and pH. Based on adsorption/desorption
studies in a variety of soils (see Table 7), triclopyr and TCP were weakly sorbed
in these soils. However, the slopes of adsorption isotherms for triclopyr and
TCP were highly nonlinear, with $1/n$ values considerably <1, indicating strong
hysteresis or resistance to leaching as soil concentrations decreased with time
(Wolt 1998).

B. Leaching

Laboratory and field studies, carried out both in Canada and in the United
States, indicate that movement of triclopyr in soil is determined by several factors such as soil organic matter content, surface cover, and pH. Its physical
properties and the field data suggest a limited potential for triclopyr to leach.

Laboratory Studies.   Lee et al. (1986) used soil columns to compare the leaching of triclopyr BEE in sand and in an high organic matter (34%) forest soil.
Following placement of a triclopyr BEE-treated soil layer at the top of each

Table 7. Soil adsorption characteristics of triclopyr BEE, triclopyr and its metabolite trichloropyridinol (TCP).

Compound	Soil type	Organic carbon (%)	pH	K_d (L kg⁻¹)	K_{oc} (L kg⁻¹)	Reference
Triclopyr BEE	Catlin silt loam	2.06	6.5	13.2	640	McCall et al. (1988)
	Commerce loam	0.58	7.5	10.3	1780	
	Tracy sandy loam	1.27	6.5	10.9	860	
	Hollis silt loam	4.91	6.1	68.0	1650	
	Average	2.2		25.6	1233	
Triclopyr	Kalkaska sand	0.73	5.0	0.98	134	Woodburn et al. (1988)
	Commerce silt loam	0.67	7.7	0.17	25	
	Mahoun clay loam	1.38	6.6	0.73	53	
	Londo sandy loam	2.25	7.5	0.57	25	
	Average	1.3		0.61	59	
	Silty loam	0.9	5.1	1.41	—	Johnson and Lavy (1994)
	Calculated	0.73	5.0	0.69	—	Wolt (1998)
		2.25	7.5	0.27	—	
		0.67	7.7	0.06	—	
		1.38	6.6	0.31	—	
	Average	1.3		0.33	—	
TCP	Immokalee sand	0.22	7.0	0.53	242	Racke and Lubinski (1992)
	Barnes clay loam	2.52	7.8	1.95	77	
	Cecil sandy loam	0.31	7.1	0.60	194	
	Catlin silt loam	2.08	6.9	1.69	81	
	Average	1.3		1.2	149	
	26 soils					
	Average[a]	1.7	—	2.9	159	Racke (1993)
TMP	5 soils					
	Average[b]	5.6	—	37	918	Racke (1993)

[a]Value range: % organic carbon, 0.45%–5.9%; K_d, 0.3–20.3 L kg⁻¹; K_{oc}, 19–389 L kg⁻¹.
[b]Value range: % organic carbon, 0.62%–16.7%; K_d, 6.7–78 L kg⁻¹; K_{oc}, 467–1402 L kg⁻¹.

column (equivalent to 5.6 kg a.e. ha^{-1}) and regular leaching (25 mm) of each column every second day over a 54-d period, soil columns were sectioned and analyzed for triclopyr and its metabolites TCP and TMP. Residues of triclopyr and its two metabolites were found in the top 10-cm layer, with total mass accounting for 65% of the initial amount applied to the column (present as 5.5% triclopyr, 88% TCP, and 6.5% TMP). No residues of either the ester or the acid were detected in any of the column eluates. Minimal leaching of triclopyr and its metabolites was attributed to the high organic matter content and low pH of the forest loam soil. In contrast, recovery from the eluates from the sand columns (as triclopyr) was quantitative after 34 d.

Using soil TLC techniques, Jotcham et al. (1989) compared the relative mobilities of triclopyr, 2,4-D, and picloram in 4 Ontario soils, 3 from forest sites with high organic matter content. Average R_f values for 4 soils were 48, 51, and 65 for 2,4-D, triclopyr, and picloram, respectively, indicating that the mobility of triclopyr was similar to that of 2,4-D and that both herbicides were less mobile in soil than picloram. This difference in mobility was attributed to the magnitude of their respective adsorption coefficients in these soils.

Field Studies.   Results of several field studies carried out under a variety of application conditions, in both Canada and the U.S., are summarized in Table 8.

Norris et al. (1987) applied triclopyr IPA at 3.4 and 10.1 kg a.e. ha^{-1} to two hillside pasture sites in western Oregon. Triclopyr residues at both sites remained in the top 30 cm of soil, with negligible or no residues detected below that depth 1 yr after the applications. Both metabolites (TCP and TMP) were also detected mainly in the 0- to 30-cm soil layer with TMP detected in the smallest concentrations. Similar mobility of triclopyr was reported by Stephenson et al. (1990), following application of triclopyr BEE at 3.0 kg a.e. ha^{-1} to sandy and clay soils in a northern Ontario forest area, both with varying amounts of vegetation cover. Even after a number of significant rainfalls from mid-June through August, about 90% or more of the triclopyr residues was present in the upper organic (vegetation) layers at both sites, with 97% being recovered within the top 15 cm of soil. Residues of triclopyr at 25- to 30-cm depths, when present, were negligible and never exceeded 6 μg kg^{-1}.

In a more extensive study, Fontaine (1990) aerially applied triclopyr BEE at 3.84 kg a.e. ha^{-1} to 100 ha of forest land (aspen and poplar), in northern Ontario, as a fall application. Duplicate soil core samples to 90-cm depth were taken at regular intervals over a period of 1 yr at 2 sites, 1 bare and the other covered by litter. The site received 89 mm of rainfall during the 30 d following application, with 41 mm occurring within 24 hr. Following snow cover in the winter, average monthly rainfall for the period May to August ranged from 56 to 157 mm the following year. Triclopyr residues at both sites were estimated to have moved to a maximum depth of 45–60 cm within 2 wk after application, after which residues continuously declined to concentrations below the limit of quantification (<0.010 mg kg^{-1}) below a depth of 30 cm.

Table 8. Soil leaching data for triclopyr under various application conditions.

Location	Land cover	Formulation/rate (kg a.e. ha^{-1})[a]	Maximum leaching (cm)	Reference
Oregon	Hillside pasture	Triclopyr IPA (3.4, 10.1)	30	Norris et al. (1987)
Ontario	Forest	Triclopyr BEE (3.0)	15	Stephenson et al. (1990)
Ontario	Forest	Triclopyr BEE (3.84)	45–60	Fontaine (1990)
Nova Scotia	Mowed hay	Triclopyr BEE (3.7)	45	Wolt et al. (1991)
North Carolina	Bare soil	Triclopyr BEE (9.07)	35–45	Petty and Gardner (1993)
California	Bare soil and grass	Triclopyr BEE (7.2)	15	Buttler et al. (1993)
Washington	Clear-cut timberland	Triclopyr BEE (6.72)	60–76	Cryer et al. (1993)
Arkansas	Flooded rice fields	Triclopyr TEA (0.41)	30.5 (TCP)	Poletika and Phillips (1996)
Louisiana	Flooded rice fields	Triclopyr TEA (0.41)	23 (TCP)	Poletika and Phillips (1996)

[a]a.e., acid equivalent.

In another Canadian study, Wolt et al. (1991) applied triclopyr BEE at 3.7 kg a.e. ha^{-1} to a freshly mowed hay field (sandy loam, 5.0%–0.6% organic matter from 0- to 90-cm soil depth) in Nova Scotia, from which the cut vegetation (red clover, timothy, and fescue) had been removed. Monitoring for residues of triclopyr and its metabolites (TCP, TMP) to a depth of 120 cm in the soil profile was carried out for more than 400 d. Triclopyr residues were primarily restricted to the organic matter layer and in the upper 45 cm of the soil profile, with occasional, inconsistent detections at lower depths during the first 28 d. Residues of the two metabolites were detected in the upper organic layer and top 15 cm of the soil profile only (limits of quantification, 0.010 mg kg^{-1} for triclopyr and TMP and 0.050 mg kg^{-1} for TCP).

Petty and Gardner (1993) applied triclopyr BEE at 9.07 kg a.e. ha^{-1} to a bare sandy clay loam soil in North Carolina. Neither triclopyr or its metabolites (TCP and TMP) were detected at depths deeper than 45 cm during the subsequent 52-wk sampling period (limits of quantification, 0.020 mg kg^{-1} for triclopyr and 0.015 mg kg^{-1} for TCP and TMP). In California, Buttler et al. (1993) also applied triclopyr BEE (7.2 kg a.e. ha^{-1}) to bare soil (loam), as well as to a native short grass-covered plot. During subsequent monitoring to 90 cm during a 62-wk period, residues of triclopyr and its metabolites (TCP and TMP) were seldom detected below the 15-cm soil depth at both sites (limits of quantification, 0.010 mg kg^{-1} for triclopyr and TMP and 0.050 mg kg^{-1} for TCP).

Cryer et al. (1993) applied triclopyr BEE at a rate of 6.72 kg a.e. ha^{-1} by helicopter to litter-covered clear-cut timberland in southwestern Washington. The upper 15 cm of soil was characterized as a loam, whereas the 15- to 120-cm profile was a clay loam. Subsequent monitoring of the site for 13 mon indicated that residues of triclopyr and its metabolites (TCP and TMP) were mainly confined to the top 15 cm of the soil profile, with only occasional detections in the 15- to 30-cm and 30- to 45-cm depths. Following application to an "exposed" site from which the litter was removed to simulate a worst case scenario for a forest ecosystem, leaching of triclopyr and its metabolites increased with time with sporadic detections of triclopyr through to the 60- to 75-cm depth. At 9–13 mon after application, TCP was generally found throughout the profile. It was suggested that, for the litter-covered sites, the litter layer retained most of the parent herbicide, which resulted in its enhanced degradation and, in turn, accounted for the decreased leaching.

Poletika and Phillips (1996) applied preflood and postflood applications of triclopyr TEA (0.41 kg a.e. ha^{-1}) to rice fields in Arkansas and Louisiana. Monitoring the sites over a 1-yr period indicated that residues of triclopyr and its metabolites (TCP and TMP) remained primarily in the upper 15 cm of soil. The maximum depth to which TCP was detected was 30 cm. At lower depths, concentrations of triclopyr and its metabolites were less than the limit of quantification (0.010 μg kg^{-1}). Concentrations of TMP were generally at least an order of magnitude less than those of TCP. Leaching of triclopyr and its metabolites was not extensive because of the limited saturated hydraulic conductivity of these rice-growing soils.

Leaching Models. Pesticide mobility and the potential of pesticides to contaminate groundwater are generally evaluated in light of several key parameters, including solubility in water, K_d, a measure of adsorption to soil, K_{oc}, a measure of adsorption to soil organic matter or soil carbon, K_H, a measure of tendency to partition between water and air, and field half-life, a measure of biotic and abiotic dissipation in field soils. The models, which utilize these parameters, are useful in comparing the relative mobility of pesticides in soil profiles.

Laskowski et al. (1982) calculated leaching index (LI) values, derived from water solubilities, field half-lives, vapor pressures, and soil adsorption constants (K_{oc}), to rank 27 pesticides in descending order of their potential for contamination of groundwater. More recently, Warren and Weber (1994) and Weber (1995) have developed another ranking called pesticide leaching potential (PLP), which is based on the field half-life, K_{oc} value, the fraction of pesticide reaching the soil during application, and application rate of a pesticide. In both rankings, triclopyr was considered less mobile than dicamba and picloram, two relatively mobile herbicides likely to be used under similar situations (Table 9).

Recently, Wolt (1998) utilized the pesticide root zone model (PRZM) for determining the soil mobility potential of triclopyr. The model develops an integrated assessment considering molecular properties, physical environment, and worst case scenarios. As a worst case scenario, Wolt (1998) considered a sandy soil profile with low organic matter content and low sorption characteristics (Table 10). In this scenario, the herbicide was applied to bare soil, and 4 d later there was sufficient rainfall (94 mm over a 6-d period) to leach water through the soil profile. During the subsequent 17-mon period, there was a total of 191 cm of rain. Half-lives for triclopyr and TCP were assumed to be 40 and 350 d, respectively, and, to account for declining microbial activity with depth, were scaled by threefold through the soil profile. Using these worst case scenario parameters, the simulation results predicted a total triclopyr flux past the root

Table 9. Comparison of leaching index (LI) and pesticide leaching potential (PLP) of triclopyr with those of dicamba and picloram, herbicides also likely to be used in pastures and noncropland areas.

Herbicide	LI values[a]	Application	PLP values[b]
Triclopyr	0.5×10^9	Foliar; noncrop[c]	7
Dicamba	11.1×10^9	Foliar; noncrop	16
Picloram	8.8×10^9	Foliar; noncrop	21
		Soil; noncrop	41

[a]Laskowski et al. (1982)
[b]Warren and Weber (1994); Weber (1995); 0, no leaching potential; 100, maximum leaching potential.
[c]noncrop: noncrop application rates are generally much higher than those used in cropland.

Table 10. Leaching assessment of triclopyr using worst case scenario in a loamy sand, with soil and rainfall characteristics used in the pesticide root zone model (PRZM) simulations/modeling study.

Number of soil sections	Soil depth range (cm)	Organic matter range (%)	Sand/clay (%)	Adsorption coefficient range (K_d)	Rainfall range over 17 mon (mm/mon)
9	0–270	1.10–0.10[a]	89.7–95.5/2.7–5.0	0.52–0.62	30.5–296[b]

[a]Decreasing with depth.
[b]Corrected for inputs such as losses in runoff (surface runoff, erosional losses, inter/base flows, etc.).

zone (0- to 1-m depth) of only 0.01% of the applied mass, and no triclopyr flux past 2.7 m. TCP, because of its longer half-life, exhibited greater leaching potential. The simulation results predicted TCP flux past the root zone equivalent to 8.7% of applied triclopyr, and 0.1% of applied triclopyr past 2.7 m.

C. Surface Runoff

Quality of the surface waters can be affected by entry of pesticides and sediments in surface runoff water. In an extensive literature review of runoff studies involving agricultural soils, Wauchope (1978) concluded that, in most instances, total pesticide runoff losses were ≤0.5% of the amounts applied, except when a major rainfall event occurred within days of application. Only a few quantitative runoff studies are available for triclopyr, and these suggest its losses in runoff are also minimal. After an early spring aerial application of triclopyr BEE at 4 kg a.e. ha^{-1} to steep hillside bush and pasture land traversed by a stream, maximum runoff losses to the stream (2.2% of amount applied) occurred 41–46 d after application following the first significant rainfall (~200 mm) in the area (Wilcock et al. 1991). Despite several subsequent significant rainfall events at the study site, weekly water samples, collected from the stream using flow-proportional water samplers, had no detectable residues of triclopyr [limit of detection (LOD), 0.1 μg L^{-1}].

Stephenson et al. (1990) followed lateral movement of triclopyr on two soil types, a sandy soil (slope 8°) and a clay soil (slope 7°), following aerial application of triclopyr BEE at 3.0 kg a.e. ha^{-1} to a northern Ontario forest. Triclopyr residues in soil, downslope from the sprayed strips, remained below the limit of quantification (0.54 μg kg^{-1}) throughout the study period, ~1 yr. Triclopyr concentrations in runoff water samples collected in trenches 12–13 m downslope of the sprayed strips were <1 μg L^{-1} 1–105 d after treatment. Although losses to the trenches could not be quantified, the presence of the herbicide in water in the trenches confirmed its transport in rainfall runoff.

D. Volatilization

Volatilization can be an important pathway for the dissipation of pesticides from treated surfaces and for their subsequent transport in the atmosphere. As a general rule, the higher the vapor pressure of a pesticide, the more likely it will volatilize from treated surfaces and, if not subsequently degraded, be detected in the atmosphere. However, vapor losses to the atmosphere are also determined, in part, by the water solubility of a pesticide, which plays a role in determining the partitioning of a pesticide between a water surface and the air above it, as defined by its Henry's law constant. In addition, meteorological factors such as temperature and windspeed, can have significant impact on these interactions.

The vapor pressures of triclopyr BEE, triclopyr, and triclopyr TEA, in decreasing order of magnitude, are 0.48, 0.17, and <0.13 mPa, respectively (see Table 1). Considering the relatively low vapor pressures of triclopyr and its ester, the rapid hydrolysis of triclopyr BEE in water or moist soil to triclopyr,

the short field half-life of triclopyr, and that, in moist soil and aquatic environments, triclopyr is most likely to predominate as an anion due to its relatively
low pK_a (2.93) value, it is unlikely that volatilization will be an important pathway for their dissipation in the environment. The low vapor pressure and high
water solubility of triclopyr TEA also point to insignificant losses of the amine
salt to the atmosphere via volatilization.

V. Persistence
A. Water and Sediments

Several field studies have established that dissipation of triclopyr in natural
waters, both static and flowing, and sediments is rapid with half-lives being less
than 4 d. Its rapid dissipation in water and degradation in sediments were attributed to photodecomposition and microbial activity, respectively.

Ponds/Lakes/Wetlands. Solomon et al. (1988) surface-sprayed triclopyr BEE
at 3.0 and 0.3 kg a.e. ha^{-1} to established limnocorrals in a northern Ontario lake
and followed its dissipation in water and sediments at regular intervals over
several months. The estimated half-lives for total triclopyr (ester plus acid) in
water at the two application rates were 4.3 and 3.8 d, respectively, with more
than 95% being dissipated within 15 d. No triclopyr residues were detected in
water after 42 d. Initial triclopyr residues in sediments were variable and were
generally more than 100 fold lower than the amounts found in the water.

Kreutzweiser et al. (1995), using large lake enclosures, reported that the half-
life for triclopyr BEE in lake water ranged from 4 to 8 d, with 78%–90%
dissipated after 15 d. Triclopyr concentrations increased linearly with time, corresponding to the linear dissipation of the ester.

Flowing Water. Several studies have shown that concentrations of triclopyr in
waters and associated sediments in streams oversprayed with triclopyr BEE or
triclopyr TEA were short lived or transient, presumably because of rapid dilution and photolysis.

Getsinger and Westerdahl (1984) treated two 0.4-ha plots in the Spring
Creek arm of Lake Seminole, GA, with triclopyr BEE, which resulted in
triclopyr concentrations of 28 and 166 µg L^{-1} in the water 1 d after treatment.
Triclopyr concentrations declined to <10 µg L^{-1} by day 7 and day 14 in the
respective plots. In both plots, triclopyr residues in the sediments remained below detection limit (<100 µg kg^{-1}) throughout the sampling period of 56 d.

Wan (1987) applied an aerial application of triclopyr BEE at 0.91 kg a.e.
ha^{-1} to about 21 ha of forest area in the Chilliwack River valley in British
Columbia, which included overspraying a flowing stream within the treated
area. Concentrations of total triclopyr (acid plus ester) in the stream water
reached a peak concentration of 620 µg L^{-1} within 0.5 hr of the commencement
of spraying but had decreased to 190 µg L^{-1} by 4 hr after spraying. Total triclopyr concentrations further decreased to 6 µg L^{-1} within 24 hr and, subsequently,

remained near the detection limit (1 μg L^{-1}) during the next 2 mon. TCP was not detected in the stream water until the second day after application, when it was present at 70 μg L^{-1}. The metabolite was continuously detected in similar concentrations over the remainder of the sampling period (59 d). Concentrations of triclopyr and TCP in the stream water may reflect, to some extent, concentrations of these compounds in the stream sediments. Triclopyr was detected at 140 μg kg^{-1} in the steam sediments 4 hr after the aerial application, with concentrations decreasing to 40 μg kg^{-1} before the first rainfall, which occurred on day 7. Residues in sediments subsequently ranged from 40 to 120 μg kg^{-1} over the next 2 mon and then decreased to near or below detection limit (20 μg kg^{-1}) thereafter. TCP was detected in the stream sediments 4 hr after application at 200 μg kg^{-1}. Over the next 2 mon, metabolite concentrations ranged from 30 to 70 μg kg^{-1}.

In a second Canadian study, initial concentrations of triclopyr BEE in stream water (Dora Creek), following direct aerial application at 3.67 kg ha^{-1} to 88 ha of the surrounding boreal forest in Ontario, ranged from 230 to 350 μg L^{-1} and then decreased to 50–110 μg L^{-1} during the first 12–14 hr after application (Thompson et al. 1991). Concentrations then declined to below the detection limit of 1 μg L^{-1} within 72 hr postapplication. Due to hydrolysis of the ester, triclopyr was also detected in the stream water, but in concentrations lower than those of the ester (maximum concentration, 140 μg L^{-1}). At 3 d postapplication, no quantifiable concentrations (>10 μg L^{-1}) of triclopyr were detected in the stream water. TCP concentrations did not exceed 10 μg L^{-1} in the stream water at any time. Concentrations of triclopyr BEE, triclopyr, and TCP in all sediment samples, except one, were below limits of quantification of 10, 100, and 50 μg kg^{-1}, respectively.

Woodburn et al. (1993b) applied triclopyr TEA at 2500 μg a.e. L^{-1} to two 4-ha areas in the Spring Creek arm of Lake Seminole in Georgia, one aerially and the other by injection. Following sampling of water and sediments in the treated areas as well as in a downstream site at regular intervals, triclopyr concentrations indicated that its dissipation followed first-order kinetics. Initial water concentrations in surface and bottom water samples ranged from 1860 to 3370 μg L^{-1} for the two treatments. The residues in water decreased rapidly, with first-order half-lives of 3.3 and 3.5 d for surface and bottom waters for the aerial application and only 0.4 d for the injection treatment. The half-lives for surface application were consistent with that (4 d) reported by Solomon et al. (1988). The major metabolite, TCP, also showed rapid dissipation and, with a half-life of 1 d, <50 μg L^{-1} was detected after the 8-hr sampling. The rapid photolysis half-life of TCP (<1 hr) under midsummer sunlight at 40° N latitude was reported earlier under laboratory conditions (Dilling et al. 1984). Triclopyr residues in sediments from both treatments were in the range of 100–640 μg kg^{-1} immediately after application. Residues in sediments were detected sporadically in samples taken on days 1 and 3, and were less than the detection limit (<100 μg kg^{-1}) thereafter. No TCP was detected in any sediment samples.

A point-source application of triclopyr BEE at two injection sites on Icewater

Creek in Ontario resulted in a maximum concentration of 848 μg L^{-1} of the ester in the stream water (Thompson et al. 1995). The decrease in BEE concentration was accompanied by rapid hydrolysis of the ester, as indicated by the presence of primarily triclopyr at 330 μg L^{-1} in water samples collected about 200 m downstream of the second injection site, 120–150 min postinjection. At downstream distances of 300 and 400 m, hydrolysis of the ester was complete and all the herbicide was present as triclopyr. Direct injection treatment resulted in sediment residues of triclopyr of 26–70 μg kg^{-1} within hours after application, depending on proximity to the injection site.

Triclopyr BEE at 3 L ha^{-1} was aerially applied by helicopter to an 8-km reach of the Ahuriri River upstream of Lake Benmore in New Zealand for control of Russell lupines (Maloney 1995). The total area of the river surface treated was 61.5 ha. Triclopyr concentrations in water samples taken over the next 6 hr at the downstream end of the treated reach ranged from 1.0 to 3.4 μg L^{-1}. Triclopyr was not detected in the river water samples subsequently collected for a period of 25 d after application.

Getsinger et al. (1996) injected triclopyr TEA 30–60 cm below the surface of the Pend Oreille River in eastern Washington State to give an average triclopyr concentration of 4.59 ±1.46 mg L^{-1} at 1 hr after application in a 4-ha treated area. Average triclopyr residues decreased to below the detection limit (HPLC analysis; 10 μg L^{-1}) in all sampling zones of the treated area by day 7. Average aqueous half-life of triclopyr throughout the treated area was calculated to be 19.4 hr.

Groundwater. There is some evidence for the occurrence of triclopyr in groundwater. Of 379 wells monitored for various pesticides, including triclopyr, in four states (ME, TX, VA, and VT) during 1984–1990, five wells in two states contained detectable triclopyr concentrations. One well in Texas and four wells in Virginia contained 0.58 and 0.006 to 0.018 μg L^{-1} of triclopyr, respectively (Anonymous 1992).

Long (1988) monitored several pesticides, including triclopyr, in groundwater near or beneath agrochemical mixing and loading facilities in Illinois. Of the 56 sites monitored, none was found to have quantifiable concentrations of triclopyr in the groundwater. However, information, regarding the depth of wells and the proportion of mixing and loading facilities at which triclopyr was used, was not provided.

Bush et al. (1988) carried out weekly monitoring of groundwater samples for triclopyr and its ester over an 18-wk period, following separate applications of triclopyr BEE and triclopyr TEA to two 100- by 500-m watersheds of poorly drained Pomona sand near Gainesville, FL. Rainfalls occurred 2 d (9 mm) and 38 d (42 mm) after application. Groundwater samples were collected from seven piezometers installed at each treated site into a shallow (1–1.5-m depth) unconfined aquifer. None of the groundwater samples contained detectable concentrations of either triclopyr or its ester (limits of detection, 0.7 and 1.0 μg L^{-1}, respectively).

B. Soil

A number of laboratory and field dissipation studies, using ester and amine salt formulations of triclopyr and carried out under a variety of use conditions, indicate that the soil half-life of triclopyr was quite variable, ranging from 8 to 96 d. The soil half-life of triclopyr BEE, however, was short, being less than a few days.

Laboratory Studies.   As early as 1977, Bidlack et al. reported that half-lives of triclopyr in silty clay loam and silt loam soils under aerobic conditions in the laboratory were 8 and 18 d, respectively. The metabolite TCP was identified and reported not to be persistent.

Johnson and Lavy (1994) monitored triclopyr-treated soil contained in Mason jars and buried at various depths in a flooded rice field for up to 3 yr. Soil half-lives of triclopyr were 10, 10, and 39 d at the 2-, 20-, and 60-cm depths, respectively. The increased half-life at the 60-cm depth was attributed to higher adsorption of triclopyr at 60 cm ($K_d = 2.75$ L kg^{-1}) compared to that at 2- and 20-cm depths ($K_d = 1.60$ and 1.41 L kg^{-1}, respectively).

Field Studies.   Several field persistence studies, using triclopyr under a variety of application situations, have been carried out in Canada and the U.S. In these studies, the soil half-lives of triclopyr varied from a few days to as long as 96 d with an average of ~36 d ($n = 12$) (Table 11).

Norris et al. (1987) applied triclopyr IPE to two hill pasture soils in western Oregon (OR), one at 3.4 and the other at 10.1 kg a.e. ha^{-1}. Using first-order kinetics, the soil half-lives of triclopyr at the two sites were 75 and 81 d, respectively. The slow dissipation of triclopyr in these pasture soils was attributed to the dry Oregon summer following the treatments.

Newton et al. (1990) applied triclopyr BEE (1.65 and 3.3 kg a.e. ha^{-1}) and TEA (2.2 and 4.4 kg a.e. ha^{-1}) by helicopter to mountainous southwest Oregon brush fields on clay loam soils. Initial soil samples, at 15-cm intervals, were collected on day 37 after application, before which no rainfall had occurred. By day 79, integrated concentrations of triclopyr in the top 60 cm of soil treated with the ester had declined to 10%–20 % of the initial concentrations detected at day 37. By 325 d, concentrations had decreased to 5%–10% of initial values. The dissipation of triclopyr residues in amine-treated plots, however, was much slower.

Stephenson et al. (1990) reported that triclopyr BEE was moderately persistent in sandy and clay soils following application at 3.0 kg a.e. ha^{-1} to a northern Ontario (ON) forested area, with 50% and 90% of the residues disappearing within 14 and 48 d, respectively. When triclopyr BEE was applied at 7.17 kg a.e. ha^{-1} to a native short grass-covered plot and to a bare loam soil in California (CA), the reported half-lives of triclopyr in the upper 0- to 15-cm layers were 40 and 39 d, respectively (Buttler et al. 1993). A somewhat longer half-life was reported by Fontaine (1990), who aerially applied triclopyr BEE at 3.84 kg a.e. ha^{-1} as a fall application to 100 ha of forest land (aspen and poplar) in northern Ontario. The half-life for total triclopyr (ester and acid) in the upper soil layers

Table 11. Persistence data for various triclopyr formulations under field conditions at sites in Canada and in the U.S.

State/province (soil cover)	Field rate (kg ha⁻¹)	Soil texture[a]	Soil organic matter (%)	Soil pH	Soil half-life[b] (d)	Reference
Field studies (triclopyr BEE) (noncrop):						
NS (mowed hay)	3.70	sl	2.87	6.40	25	Wolt et al. (1991)
CA (grass cover)	7.17	l	2.41	5.60	40	Buttler et al. (1993)
CA (bare soil)	7.17	l	2.41	5.60	39	Buttler et al. (1993)
NC (bare soil)	9.07	sl	0.66	6.30	11	Petty and Gardner (1993)
Field studies (triclopyr BEE) (forest land):						
ON (litter)	3.00	s/c	5.2/3.4	4/5	14	Stephenson et al. (1990)
ON (litter)	3.84	fsl-sic	1.8/3.3	6.4/6.9	26	Fontaine (1990)
WA (litter)	6.72	l	2.12	5.50	37	Cryer et al. (1993)
WA (bare)	6.72	l	2.12	5.50	96	Cryer et al. (1993)
Field studies (triclopyr IPA) (pasture):						
OR (grazed)	3.40	—	—	—	75	Norris et al. (1987)
	10.1	—	—	—	81	
Flooded (triclopyr TEA[1]) (rice)						
AR (preflooded)	0.41	sic	0.56	5.10	8	Poletika and Phillips (1996)
(postflooded)					2	
LA (preflooded)	0.41	sicl	1.26	5.60	2	
(postflooded)					3	

NS, Nova Scotia; ON, Ontario, Canada

AR, Arkansas; CA, California; LA, Louisiana, OR, Oregon; WA, Washington, U.S.

[a]sicl, silty clay loam; sil, silty loam; sl, sandy loam; l, loam; fsl, fine sandy loam; sic, silty clay.

[b]Average soil half-life = 44.4 d, using all dryland data ($n = 10$).

was reported as 62 d when data for the whole year, which included the long winter period, were included. This is a more realistic estimate of the half-life for total triclopyr in soil than that (26 d) calculated when only the first 91 d after application were considered. There was no difference in half-life between bare soil and soil covered with litter.

Wolt et al. (1991) applied triclopyr BEE at 3.7 kg a.e. ha^{-1} to a sandy loam soil, with freshly mowed hay (fescue, red clover, timothy) cover in Nova Scotia (NS). Monitoring of the treated site for both triclopyr and its metabolites (TCP and TMP) for more than 400 d showed that the apparent half-life for triclopyr degradation in the upper 45 cm of soil was only 15 d, whereas that for total mineralization of the parent and its metabolites was 25 d. An application of triclopyr BEE at 9.07 kg a.e. ha^{-1} to a bare sandy loam soil in North Carolina (NC) resulted in rapid degradation of the ester to the acid with a half-life of 1.4 d in the upper 0- to 7.5-cm of soil (Petty and Gardner 1993). Total dissipation of the ester plus the acid was also rapid, the total triclopyr half-life being 10.6 d. Cryer et al. (1993) applied triclopyr BEE by helicopter to litter-covered clear-cut timberland in southwest Washington (WA) at a rate of 6.72 kg a.e. ha^{-1}. The estimated soil half-lives for total triclopyr (ester plus acid) residues were 96 d in exposed (organic layer removed) soil and 37 d in unexposed soil.

A dual application (pre- and postflood) of triclopyr TEA at 0.41 kg a.e. ha^{-1} to rice fields in Arkansas (AR) and Louisiana (LA) resulted in its rapid dissipation under this aquatic food crop use (Poletika and Phillips 1996). The half-lives of triclopyr for preflood and postflood applications were 7.6 and 1.8 d in Arkansas and 2.2 and 3.4 d in Louisiana, respectively. The corresponding half-lives for TCP were 1.0 and 0.3 d for Arkansas and 1.2 and 0.7 d for Louisiana, whereas those for TMP were 0.4 and 4.1 d for Arkansas and 0.4 and 2.7 d for Louisiana. It was suggested that the rapid dissipation of triclopyr and TCP under flooded conditions was primarily caused by photodegradation.

In summary, triclopyr persistence in soil under ideal laboratory conditions was very short, with half-lives of only 8–18 d. Under field conditions, triclopyr BEE hydrolyzed rapidly to the acid form with an half-life of 1–2 d. The field persistence of triclopyr (acid) was, however, quite variable, ranging from only 3–8 d in flooded rice fields in Arkansas and Louisiana to as high as 96 d in pastures in western Oregon and Washington. Under moist forest canopies, its half-life ranged from 11 to 50 d.

VI. Environmental Exposure Assessment
A. Surface Water

Laboratory and field data indicate that, due to rapid hydrolysis, the aquatic field half-life of triclopyr BEE in surface waters is <2 d, whereas that of the resulting triclopyr is significantly longer (<12 d). In flowing waters, the occurrence of triclopyr is considered transient, most likely, in large measure, due to dilution. Thus, triclopyr, regardless of whether it is formulated as an ester or amine salt, is considered to be relatively nonpersistent in surface waters.

B. Groundwater

The Environmental Fate and Effects Division (EFED) of the Office of Pesticides Programs (OPP), a component of the U.S. Environmental Protection Agency (USEPA), regularly conducts risk assessments of pesticide use, including surface water and groundwater contamination. The guidelines for assessing environmental risk resulting from surface water and groundwater contamination are based, in large measure, on the following environmental measurements (indicated in Table 12):

a. Field and laboratory studies on persistence
b. Relative mobility characteristics of the pesticide
c. Movement to 70–90 cm depth through soil profile
d. Reported detections from groundwater monitoring studies

Noncompliance with these guidelines may trigger restricted use or groundwater label advisories, depending on the end use of the water being contaminated.

In response to an EFED assessment of triclopyr use that raised concerns concerning triclopyr contamination of surface water and groundwater, Wolt (1998), using environmental fate data and worst case use scenarios, carried out an evaluation to address these concerns regarding water resources in situations in which triclopyr BEE or TEA was used in accordance with label rates and application practices. Using guidelines established by the EFED, available field data were reviewed and, using models where feasible, evidence from an integrated risk assessment was offered to mitigate concerns relative to surface water and groundwater resources contamination with triclopyr. Triclopyr compliance with EFED guidelines, based on evaluation of available environmental data for these guidelines, has been summarized in Table 12.

The following points highlight the environmental data for triclopyr mobility and persistence and its potential to contaminate groundwater in light of the EFED guidelines:

1. Triclopyr soil half-lives are highly variable (2–96 d) and on average ~5 wk, somewhat greater than the stated EFED guideline of 3 wk. Availability of soil moisture is critical for its degradation to TCP, its primary metabolite. Soil degradation of TCP is microbial, and it can be readily induced.
2. Adsorptive characteristics of triclopyr, together with the Tier II PRZM modeling for a vulnerable site, indicate limited leaching potential of triclopyr. This result is corroborated by field leaching data for a variety of soils and environmental conditions in which leaching ranged from <15 to 76 cm, with an average of 39 cm, well below the EFED guideline of 75 cm.
3. Lack of significant number of hits for triclopyr in a number of groundwater monitoring studies further attests to its limited potential to contaminate groundwater. In a major monitoring study in the U.S., only 5 samples from 2 states, of 379 samples collected from 4 states, contained triclopyr residues, again well below the EFED guidelines. In another study, none of the 59 handling/mixing sites in 1 state contained triclopyr residues in their groundwater.

Table 12. Triclopyr compliance versus USEPA-OPP Environmental Fate and Effects Division (EFED) guidelines.

Environmental factors	Criteria	EFED guideline	Triclopyr behavior	Reference
Persistence	Field half-life	>3 wk	75, 81 d (OR, USA)	Norris et al. (1987)
			26 d (ON, Canada)	Fontaine (1990)
			14 d (ON, Canada)	Stephenson et al. (1990)
			25 d (ON, Canada)	Wolt et al. (1991)
			39, 40 d (CA, USA)	Buttler et al. (1993)
			11 d (NC, USA)	Petty and Gardner (1993)
			37, 96 d (WA, USA)	Cryer et al. (1993)
			8, 2 d (LA, USA)	Poletika and Phillips (1996)
			2, 3 d (AR, USA)	Poletika and Phillips (1996)
			Mean = 33 d (n = 14)	
	Laboratory half-life	>3 wk	8, 18 d	Bidlack et al. (1977)
	Hydrolysis	<10% in 30 d	>30 d	Cleveland and Holbrook (1991)
	Soil photolysis	<10% in 30 d	12 wk (sterile soil)	Wolt (1998)
Mobility	Soil adsorption (K_d)	≤5 L kg^{-1}	<1 L kg^{-1} (est.)[a]	(See Table 7)
	Soil adsorption (K_{oc})	≤500 L kg^{-1}	<100 L kg^{-1} (est.)	
	Depth of leaching, field study	>75 cm	<15 cm	Norris et al. (1987)
			<60 cm	Fontaine (1990)
			<15 cm	Stephenson et al. (1990)
			<45 cm	Wolt et al. (1991)
			<15 cm	Buttler et al. (1993)
			<76 cm	Cryer et al. (1993)
			<45 cm	Petty and Gardner (1993)
			Mean = 39 (n = 7)	

Table 12. (Continued).

Environmental factors	Criteria	EFED guideline	Triclopyr behavior	Reference
Detections in groundwater	No. sampled/no. detections	≥25 wells in four or more states	379/5 wells in two states	Anonymous (1992) (US EPA database)
	Concentration range (μg L^{-1})	<350 μg L^{-1} (est. HAL[b])	(1) 0.58 μg L^{-1} (TX, USA) (4) 0.006–0.018 μg L^{-1} (VA, USA) one county with <1% of HAL	
	Maximum detects as % of MCL[c]	>three counties show detects > 10% of MCL or HAL		
	No. sampled/no. detections	(Manufacturer/ commercial sites)	56/0	Long (1989)
	No. sampled/no. detections	(Weekly sampling, October–March)	~36/0	Bush et al. (1988)

OR, Oregon; ON, Ontario; CA, California; NC, North Carolina; WA, Washington; LA, Louisiana; AR, Arkansas; TX, Texas; VA, Virginia.
[a]Estimated
[b]Health advisory level (HAL).
[c]Maximum concentration level.

4. Maximum reported concentration of triclopyr in groundwater was 0.58 μg L^{-1}, <1% of the EFED estimated Health Advisory Level (HAL). However, this concentration exceeds 0.1 μg L^{-1}, the maximum pesticide concentration permitted in water intended for human consumption by the Council of the European Union (Anonymous 1998).

Summary

Triclopyr is nonpersistent in surface water. It has limited mobility and low to medium persistence in soil. Considering its adsorptive characteristics and that it dissipates via multiple pathways, such as photolysis, plant metabolism, and microbial degradation, its potential to leach to depth in soil and to contaminate groundwater is low. This conclusion is corroborated by field leaching and groundwater monitoring data, both derived from use areas in several states in the U.S. and sites directly near handling/mixing facilities. Even when detected in the groundwater, e.g., five reported detections in two states in the U.S., the highest concentration was well below the estimated HAL of the USEPA.

References

Anonymous (1992) Pesticides in groundwater database: a compilation of monitoring studies: 1971–1991. National summary. U.S. Environmental Protection Agency, Washington, DC.

Anonymous (1994) Herbicide Handbook, 7th Ed. Weed Science Society of America, Champaign, IL.

Anonymous (1997) The Pesticide Manual, 11th Ed. The British Crop Protection Council. Lavenham Press, Lavenham, Suffolk, U.K.

Anonymous (1998) The Council of the European Union, Council Directive 98/83/EC on the quality of water intended for human consumption. Official Journal L330, 05/12/98 p. 0032–0054.

Bidlack HD (1977) Aerobic degradation of 3,5,6-trichloro-2-pyridinol in 15 agricultural soils. GH-C 991. DowElanco, Indianapolis, IN.

Bidlack HD (1978) The hydrolysis of triclopyr EB ester in buffered deionized water, natural water, and selected soils. GH-C 1106. Dow Chemical U.S.A., Midland, MI.

Bidlack HD, Laskowski DA, Swann RL, Comeaux LB, Jeffries TK (1977) Comparison of the degradation rates and decomposition products of ^{14}C-triclopyr in aerobic and waterlogged soil. GH-C 919R. DowElanco, Indianapolis, IN.

Bovey RW, Ketchersid ML, Merkle MG (1979) Distribution of triclopyr and picloram in huisache (Acacia farnesiana). Weed Sci 27:527–531.

Bush PB, Neary DG, Taylor JW (1988) Effect of triclopyr amine and ester formulations on groundwater and surface water quality in the coastal plain. Proc South Weed Sci Soc 41:226–232.

Buttler IW, Roberts DW, Siders LE, Gardner RC (1993) Non-crop right-of-way terrestrial field dissipation of triclopyr in California. GH-C 3007. DowElanco, Indianapolis, IN.

Byrd BC, Wright WG, Warren LE (1975) Vegetation control with Dowco® 233 herbicide. Proc West Soc Weed Sci 28:44–48.

Cessna AJ, Grover R (1978) Spectrophotometric determination of dissociation constants of selected acidic herbicides. J Agric Food Chem 26:289–292.

Cleveland CB, Holbrook DL (1991) A hydrolysis study of triclopyr. GH-C 2491. Dow Elanco, Indianapolis, IN.

Cryer SA, Cooley T, Dixon-White H, Schuster L (1993) The dissipation and movement of triclopyr in a northern USA forest system. GH-C 3152. DowElanco, Indianapolis, IN.

Cserjesi AJ, Johnson EL (1972) Methylation of pentachlorophenol by *Trichoderma virgatum*. Can J Microbiol 18:45–49.

Curtis RF, Land DG, Griffiths NM, Gee M, Robinson D, Peel JL, Dennis C, Gee JM (1972) 2,3,4,6-Tetrachloroanisole association with musty taint in chickens and microbial formation. Nature (Lond) 235:223–235.

Dilling WL, Lickly LC, Lickly TD, Murphy PG, McKellar RL (1984) Organic photochemistry. 19. Quantum yields for *O,O*-diethyl-*O*-(3,5,6-trichloro-2-pyridinyl) phosphorothioate (chlorpyrifos) and 3,5,6-trichloro-2-pyridinol in dilute aqueous solutions and their environmental phototransformation rates. Environ Sci Technol 18:540–543.

Feng Y (1995) Transformation of 3,5,6-trichloro-2-pyridinol, a metabolite of pyridine-based pesticides. PhD dissertation. Pennsylvania State University, University Park, PA.

Fontaine DD (1990) Dispersal and degradation of triclopyr within a Canadian boreal forest ecosystem following an aerial application of Garlon 4. GH-C 2314. Dow Chemical, Midland, MI.

Getsinger KD, Westerdahl HE (1984) Field evaluation of Garlon 3A (triclopyr) and 14-ACE-B (2,4-D BEE) for the control of Eurasian watermilfoil. Miscellaneous paper A-84-5. Waterways Experiment Station, U.S. Army Corps of Engineers, Vicksburg, MS.

Getsinger KD, Madsen JD, Netherland MD, Turner EG (1996) Field evaluation of triclopyr (Garlon 3A) for controlling Eurasian watermilfoil in the Pend Oreille River, Washington. Contract WES: Tech Rep A-96-1. NTIS/AD-A304 807/1. Waterways Experiment Station, U.S. Army Corps of Engineers, Vicksburg, MS.

Gorrell RM, Bingham SW, Foy CL (1988) Translocation and fate of dicamba, picloram and triclopyr in horsenettle, *Solanum carolinense*. Weed Sci 36:447–452.

Johnson WG, Lavy TL (1994) In-situ dissipation of benomyl, carbofuran, thiobenzcarb, and triclopyr at three soil depths. J Environ Qual 23:556–562.

Johnson WG, Lavy TL, Gbur EE (1995) Persistence of triclopyr and 2,4-D in flooded and nonflooded soils. J Environ Qual 24:493–497.

Jotcham JR, Smith DW, Stephenson GR (1989) Comparative persistence and mobility of pyridine and phenoxy herbicides in soil. Weed Technol 3:155–161.

Kreutzweiser DP, Thompson DG, Capell SS, Thomas DR, Staznik B (1995) Field evaluation of triclopyr ester toxicity to fish. Arch Environ Contam Toxicol 28:18–26.

Laskowski DA, Goring CAI, McCall PJ, Swann RL (1982) Terrestrial environment. In: Conway RA (ed) Environmental Risk Analysis for Chemicals. Von Nostrand Rheinhold, New York, pp 198–240.

Lee CH, Oloffs PC, Szeto SY (1986) Persistence, degradation, and movement of triclopyr and its ethylene glycol butyl ether ester in a forest soil. J Agric Food Chem 34:1075–1079.

Lewer P, Owen WJ (1989) Amino acid conjugation of triclopyr by soybean cell suspension cultures. Pestic Biochem Physiol 33:249–256.

Lewer P, Owen WJ (1990) Selective action of the herbicide triclopyr. Pestic Biochem Physiol 36:187–200.

Long T (1988) Groundwater contamination in the vicinity of agrichemical mixing and loading facilities. In: Proceedings of the 16th ENR Annual Conference: Pesticides and Pest Management, November 12–13, 1987, Chicago, IL, p. 133–149.

Maloney RF (1995) Effect of the herbicide triclopyr on the abundance and species composition of benthic aquatic macroinvertebrates in the Ahuriri River, New Zealand. NZ J Mar Freshw Res 29:505–515.

McCall PJ, Gavit PD (1986) Aqueous photolysis of triclopyr and its butoxyethyl ester and calculated environmental photodecomposition rates. Environ Toxicol Chem 5: 879–885.

McCall PJ, Laskowski DA, Jeffries TK (1976) Degradation of ^{14}C-triclopyr in sterile and non-sterile soil. GH-C 960. Dow Chemical Company, Midland, MI.

McCall PJ, Laskowski DA, Bidlack HD (1988) Simulation of the aquatic fate of triclopyr butoxyethyl ester and its predicted effects on Coho salmon. Environ Toxicol Chem 7:517–527.

Newton M., Roberts F, Allen A, Kelpass B, White D, Boyd P (1990) Deposition and dissipation of three herbicides in foliage, litter and soil of brush fields of southwest Oregon. J Agric Food Chem 38:574–583.

Norris LA, Montgomery ML, Warren LE (1987) Triclopyr persistence in western Oregon hill pastures. Bull Environ Contam Toxicol 39:134–141.

Petty DG, Gardner RC (1993) Right-of-way terrestrial dissipation study of triclopyr in North Carolina. GH-C 3123. DowElanco, Indianapolis, IN.

Poletika NN, Phillips AM (1996) Field dissipation of triclopyr in southern U.S. rice culture. CH-C 3894. DowElanco, Indianapolis, IN.

Racke KD (1993) Environmental fate of chlorpyrifos. Rev Environ Contam Toxicol 131: 1–150.

Racke KD, Lubinski RN (1992) Sorption of 3,5,6-trichloro-2-pyridinol in four soils. GH-C 2821. DowElanco, Indianapolis, IN.

Radosevich SR, Bayer DE (1979) Effect of temperature and photoperiod on triclopyr, picloram and 2,4,5-T translocation. Weed Sci 27:22–27.

Skurlatov YI, Zepp RL, Bagman GH (1983) Photolysis rates of (2,4,5-trichlorophenoxy) acetic acid and 4-amino-3,5,6-trichloropicolinic acid in natural waters. J Agric Food Chem 31:1065–1071.

Solomon KR, Bowhey CS, Liber K, Stephenson GR (1988) Persistence of hexazinone (Velpar), triclopyr (Garlon), and 2,4-D in a northern Ontario aquatic environment. J Agric Food Chem 36:1314–1318.

Stephenson GR, Solomon KR, Bowhey CS, Liber K (1990) Persistence, leachability, and lateral movement of triclopyr (Garlon) in selected Canadian forestry soils. J Agric Food Chem 38:584–588.

Szeto SY (1993) Determination of kinetics of hydrolysis by high-pressure liquid chromatography: application to hydrolysis of the ethylene glycol butyl ether ester of triclopyr. J Agric Food Chem 41:1118–1121.

Thompson DG, Staznik B, Fontaine DD, Mackay T, Oliver GR, Troth J (1991) Fate of triclopyr ester (Release®) in a boreal forest stream. Environ Toxicol Chem 10: 619–632.

Thompson DG, Kreutzweiser DP, Capell SS, Thomas DR, Staznik B, Viinikka T (1995) Fate and effects of triclopyr ester in a first-order forest stream. Environ Contam Toxicol 14:1307–1317.

Wan MT (1987) The persistence of triclopyr and its pyridinol metabolite in a coastal British Columbia stream. Environmental Protection Regional Program Report 86–24. Environment Canada, West Vancouver, BC.

Warren RL, Weber JB (1994) Evaluating pesticide movement in North Carolina soils. Soil Sci Soc North Carolina Proc 37:23–35.

Wauchope RD (1978) The pesticide content of surface water draining from agricultural fields: a review. J Environ Qual 7:459–472.

Wauchope RD, Butler TM, Hornsby AG, Augustine-Beckers PWM (1992) The SCS/ ARS/CES pesticide properties data base for environmental decision making. Rev Environ Contam Toxicol 123:1–155.

Weber JB (1995) Section II: Safety. Table on relative pesticide leaching potential (PLP) indices and ratings for 100 commonly used pesticides. In: North Carolina Agricultural Chemicals Manual. College of Agriculture and Life Sciences, NC State University, Raleigh, NC, pp 30–32.

Whisenant SG, McArthur ED (1989) Triclopyr persistence in northern Idaho forest vegetation. Bull Environ Contam Toxicol 42:660–665.

Wilcock RJ, Costley KJ, Cowles RJ, Wilson B, Southgate P (1991) Stream run-off losses and soil and grass residues of triclopyr applied to hillside gorse. N Z J Agric Res 34: 351–357.

Wolt JD (1998) Ground and surface water exposure assessment for triclopyr. GH-C 4350. DowElanco, Indianapolis, IN.

Wolt JD, Morgan RW, Woodburn KB (1991) Field dissipation of triclopyr in an Eastern Canadian soil following application of Garlon 4 herbicide. GH-C 2672. DowElanco, Indianapolis, IN.

Woodburn KB, Fontaine DD, Richards JF (1988) A soil adsorption/desorption study of triclopyr. GH-C 2107. DowElanco, Indianapolis, IN.

Woodburn KB, Fontaine DD, Bjerke EL (1989) Photolysis of picloram in dilute aqueous solution. Environ Toxicol Chem 8:769–775.

Woodburn KB, Batzer FR, White FH, Schultz MR (1993a) The aqueous photolysis of triclopyr. Environ Toxicol Chem 12:43–55.

Woodburn KB, Green WR, Westerdahl HE (1993b) Aquatic dissipation of triclopyr in Lake Seminole, Georgia. J Agric Food Chem 41:2172–2177.

Zepp RL, Cline DM (1977) Rates of direct photolysis in aquatic environment. Environ Sci Technol 11:359–366.

Manuscript received June 2; accepted June 7, 2001.

Rev Environ Contam Toxicol 174:49–170

Physical and Chemical Properties of Pyrethroids

Dennis A. Laskowski

Contents

Communicated by George W. Ware.

D.A. Laskowski (✉)
D.A. Laskowski Consulting, 4600 Hickory Court, Zionsville, IN 46077, USA.

I. Introduction

Agrochemicals undergo analyses of exposure concentrations and persistence to assess the potential impact of the chemicals from release into the environment. The estimates of exposure are compared with relevant toxicity data to character- ize the potential risk to classes of organisms undergoing risk analysis. Under regulatory processes in the United States and the European Union, the predic- tions of exposure are developed from exposure models of varying complexity, using the measurements of an agrochemical's physical and chemical properties as model input. It is necessary to have knowledge of these properties so that assessment of potential risk can be completed.

This review has collated, in a systematic manner, the physical and chemical property information relevant to the prediction of the environmental fate for synthetic pyrethroids. Sources of information were internal company reports submitted to the author by members of the Pyrethroid Working Group (PWG). Generally these reports had been developed by PWG member companies to address either U.S. or European environmental fate data needs for purposes of pesticide registration. Approximately 150 reports citing information on pyre- throid physical and chemical properties were examined during the preparation of this review. With two exceptions, the review does not include literature cita- tions from the open literature, and any information from such sources would be supplemental to that cited here.

This review allows comparison of physical and chemical property informa- tion among the pyrethroids. During review development, each report was evalu- ated in terms of the methodology used by an experimenter, the specific problems associated with measurement of physical and chemical properties of pyrethroids, and the derivation of endpoints such as half-lives and adsorption/desorption con- stants. This practice made it possible to distinguish between values of high and low confidence and thus avoid experimental outliers or anomalous results during the comparison and contrasting of the pyrethroid properties.

Because pyrethroids occur naturally as mixtures of stereoisomeric forms, property measurements at some times have been made with the mixtures and at other times with pure forms of specific stereoisomers. Attempts were not made to separate the information for one stereoisomeric form from another, but in- stead the data were combined into a single category for a pyrethroid regardless of the isomeric form used in an experiment. Because pyrethroids are used gener- ally as mixtures of stereoisomers, it was assumed that the combining of all stereoisomeric information into a single category serves to capture the overall properties of a pyrethroid.

What follows in this review is a summarization and discussion of information on the physical and chemical properties of nine pyrethroids—bifenthrin, cyfluthrin, cypermethrin, deltamethrin, esfenvalerate, fenpropathrin, lambda-cyhalothrin, permethrin, and tralomethrin—prepared from individual pyrethroid sections appearing later in this review. The separate sections, one for each pyrethroid, compile and summarize the physical and chemical property information derived from review of the reports provided by PWG member companies.

II. General Findings

Table 1 lists values for key environmental properties for each of the nine pyrethroids. In the table, MW is pyrethroid molecular weight, Log P is the log of the octanol–water partition coefficient K_{ow}, CLog P is a value of Log P calculated from molecular structure, VP is the vapor pressure, WS is the water solubility, K_h is the Henry's law constant, BCF is the bioconcentration factor in fish, and K_{oc} is the soil adsorption partition coefficient based on soil organic carbon.

In general, data in the table indicate that all pyrethroids are highly nonpolar chemicals of low water solubility, low volatility, and high octanol–water partition coefficients and have high affinity to soil or sediment particles. These chemicals thus have little mobility in soils and will be associated mainly with the sediments of natural water systems. These characteristics could suggest a certain propensity for pyrethroids to bioconcentrate into living organisms; however, BCF values are lower than might be expected, primarily because of pyrethroid capacity to be metabolized and thus eliminated by an organism.

In water solution, the pyrethroids are somewhat stable at acid and neutral pH but begin to hydrolyze readily under alkaline conditions. Exceptions at higher pH are bifenthrin, esfenvalerate, and permethrin, which hydrolyze slowly under alkaline conditions. Photolytically, the pyrethroids vary in their susceptibility to light. Water solutions of cyfluthrin and tralomethrin are fairly susceptible; lambda-cyhalothrin, esfenvalerate, deltamethrin, permethrin, and cypermethrin not quite as susceptible; and bifenthrin and fenpropathrin are the least susceptible. On soil, all but cypermethrin can be degraded at rates similar to photolytic rates in water. However, it is noted that cypermethrin is a special case because it was examined only in wet and not dry soil, which, as discussed later, may have had an impact on the photolysis outcome.

The degradation rate of the pyrethroids in aerobic soils varies moderately, ranging in half-life from 3.25 (tralomethrin) to 96.3 d (bifenthrin). Degradation in anaerobic soils and anaerobic aquatic systems tends to continue at rates similar to aerated soils. Rates in aerobic aquatic environments behave in a similar fashion.

In the next several sections a more detailed series of summary comments and findings are provided that add perspective to the overview just given. Physical

Table 1. Summary of physical and chemical environmental properties for pyrethroids.

Property	Bifenthrin	Cyfluthrin	Cypermethrin	Deltamethrin	Esfenvalerate	Fenpropathrin	Lambda-cyhalothrin	Permethrin	Tralomethrin
MW	422.9	434.3	416.3	505.2	419.9	349.4	449.9	391.3	665
Log P	6.40	5.97	6.54	4.53	5.62	6.00	7.00	6.10	5.05
CLog P	7.2	6.4	6.1	6.5	6.8	5.7	6.1	6.9	7.6
VP, mm Hg	1.8e-7	1.5e-8	2.5e-9	9.3e-11	1.5e-9	1.4e-8	1.6e-9	1.5e-8	1.8e-11
WS, ppm	1.4e-5	2.3e-3	4.0e-3	2.0e-4	6.0e-3	1.03e-2	5.0e-3	5.50e-3	8.4e-2
K_h, atm m^3 mol^{-1}	7.2e-3	3.7e-6	3.4e-7	3.1e-7	1.4e-7	6.3e-7	1.9e-7	1.4e-6	1.9e-10
BCF (fish)	6,090	719	597	698	2,390	359	2,240	558	
Soil adsorption, K_{oc}	237,000	124,000	310,000	704,000		42,500	326,000	277,000	295,000
Hydrolysis half-life, d:									
pH 5	Stable	Stable	619	Stable	Stable	1,670	Stable	Stable	167
pH 7	Stable	183	274	Stable	Stable	555	Stable	Stable	15.5
pH 9	Stable	1.84	1.90	2.15	Stable	14.4	8.66	242	38.9
Photolysis half-life, d:									
Water	408	0.673	30.1	55.5	17.2	603	24.5	110	2.47
Soil	96.9	5.02	165	34.7	10.0	4.47	53.7	104	3.87
Soil Deg half-life, d:									
Aerobic soil	96.3	11.5	27.6	24.2	38.6	22.3	42.6	39.5	3.25
Anaerobic soil	425	33.6	55.0	28.9	90.4	276		197	5.00
Aquat Deg half-life, d:									
Aerobic aquatic		7.44		79.5	72.3		21.9		
Anaerobic aquatic		7.4				73.6			

MW, molecular weight; VP, vapor pressure; WP, water solubility; K_n, Henry's law constant; BCF, bioconcentration factor; K_{oc}, With the exception of MW, values are averages of data from studies having highest level of experimental confidence.

properties are considered first, and these are followed by comments on the chemical properties of pyrethroids.

A. Physical Properties

The highly nonpolar nature of pyrethroids is largely responsible for their characteristic physical properties. This marked nonpolarity also is a source of challenge to experimenters working with these chemicals because of their lack of water solubility and tendency to sorb strongly to any type of surface, i.e., the surfaces of soil solids in sorption tests as well as to the surfaces of glassware used in the measurement of physical properties. This propensity makes it difficult for researchers to consistently obtain artifact-free measurements, and it has had an impact on the quality of some physical property measurements. Whenever appropriate, difficulties are presented and discussed in this chapter under the corresponding property subsections, but for more details and information the reader is encouraged to turn to the individual pyrethroid sections that serve as the source of information presented here.

1. Octanol–Water Partition Coefficient.   Perhaps the property that indicates the nonpolar nature of pyrethroids best is the octanol–water partition coefficient, K_{ow}. In Table 1, two values for K_{ow} (Log P and CLog P) are given for each chemical. Log P is a measured value extracted from the individual pyrethroid papers, and CLog P is the value calculated from pyrethroid molecular structure by Dr. Geoff Briggs (Briggs 1999) of AgrEvo, using the MedChem Software program, release 3.42. This program uses methodology developed originally by Rekker (1977) and modified later by Hansch and Leo (1979) to estimate K_{ow} from the molecular structure of a chemical.

   CLog P in addition to Log P is provided in Table 1 because researchers sometimes had difficulty in measuring Log P due to the ready formation of emulsions during the equilibration of a pyrethroid between octanol and water phases. Because of the pyrethroids' highly nonpolar nature, very large differences in concentrations exist between the water phase and octanol phase, making it difficult to prevent the contamination of the water phase with the higher concentrations of chemical in the octanol phase. The result could be abnormally high water-phase concentrations and thus artifactually low K_{ow}. Table 1 shows a trend for K_{ow} measurements to be lower than calculated values, and this may have been caused by the difficulties already cited. CLog P generally is greater than 5.7, suggesting that all pyrethroids are similarly high in nonpolarity and thus K_{ow} (assuming the calculation provides a more proper measure of K_{ow}).

2. Vapor Pressure.   Vapor pressures of pyrethroids usually were measured by some variance of the gas saturation method. Gas at known temperature and pressure is saturated with pyrethroid vapor by passage through a generator column filled with the chemical coated on a suitable column packing. The temperature for measurement may or may not have been at 20°–25°C because some studies utilized elevated temperatures and then extrapolated vapor pressure data back to 20° or 25 °C. With the minor exception of this extrapolation, the methodology appears generally adequate to provide valid vapor pressures for the pyrethroids. Table 1 indicates vapor pressures range from 1.8×10^{-7} to 1.8×10^{-11} mm Hg. Vapor pressures are not high for any of the pyrethroids.

3. *Water Solubility.*   Similar to Log P, water solubility measurements of pyrethroids can be technically challenging because of the potential for formation of pyrethroid suspensions during saturation of the water phase when a stirring technique is used to effect saturation by equilibration of water with excess chemical. Without great care, a suspension and not a true solution could be formed, thus causing apparent solubility to be higher than reality.

Approximately half the experiments used a column saturation technique that does not produce the suspension artifact, providing water solubilities usually lower than those achieved with stirring. Whenever possible, values generated by the column saturation methodology were selected to represent the pyrethroids instead of those developed by a stirring technique.

Table 1 indicates pyrethroid water solubilities are generally of the order of 0.001–0.01 ppm, although they range from 0.000014 to 0.084 ppm.

4. *Henry's Law Constant.*   Henry's law constants (K_h) are calculated from the vapor pressures and water solubilities for each pyrethroid in Table 1 by the following equation:

$$K_h = \frac{VP \times MW \times 0.001316}{WS} \qquad (1)$$

where: K_h = Henry's law constant (atm m^3 mol^{-1}), VP = vapor pressure (mm Hg), MW = molecular weight, WS = water solubility (ppm), and 0.001316 = conversion factor to attain proper units.

K_h values tend not to be high for any pyrethroid, suggesting that pyrethroids are not very volatile from water.

5. *Bioconcentration Factor.*   The tendency for pyrethroids to bioaccumulate is represented in Table 1 by their accumulation in fish exposed continuously to ^{14}C-labeled chemical in flow-through exposure systems. Bioconcentration factor (BCF) was calculated from levels of total ^{14}C measured in fish and exposure water over the course of the study. Corrections for possible metabolites in water or fish were not necessarily made, which may contribute some uncertainty to the BCF values listed in Table 1. BCF values tend to be lower than expected from the highly nonpolar nature of pyrethroids because the bioconcentration studies show that fish can metabolize pyrethroids and keep accumulation in their bodies at low levels through the continuous turnover of chemical.

6. *Soil Sorption.*   Sorption studies with pyrethroids have proved difficult because of extremely low pyrethroid water solubilities, a tendency for pyrethroids

to sorb significantly to surfaces of any equipment used in sorption experiments, and a lack of stability of parent chemical during equilibration; this provides some uncertainty to the sorption data summarized in Table 1 and then presented again in greater detail in Tables 2 through 4. To address this uncertainty, the experimental designs of all sorption experiments were evaluated and then assigned confidence ratings from 1 to 10 (10 being highest), keeping in mind the issues already listed. These ratings were used as an aid in the selection of data deemed most representative for each pyrethroid.

Many of the experiments were conducted as Freundlich sorption studies that measured sorption at a number of soil concentrations. However, some exposure models do not utilize Freundlich sorption constants (no correction for concentration effects on sorption) but instead use simple partition coefficients (K_d) or K_d that is corrected for soil organic carbon content (K_{oc}). Therefore, sorption data in this review are grouped into soil concentration categories (<0.5 ppm, 0.5 to <1 ppm, 1–5 ppm, and >5 ppm) based on sorbed soil concentrations so that, if desired or necessary, concentration effects on sorption can be included in the modeling by selection of an appropriate concentration category as dictated by the application rate of a pyrethroid.

Because most pyrethroid sorption studies were conducted at four concentrations in each of a variety of soils, a large number of adsorption and desorption partition measurements are available, making the sorption–desorption database for pyrethroids fairly extensive. This review contains 392 adsorption (K_d) and 314 desorption (K_{dd}) values for the pyrethroids overall. Table 2 summarizes the adsorption data, Table 3 the desorption data, and Table 4 the Freundlich adsorp-

Table 2. Pyrethroid soil adsorption partition coefficients.

Chemical	Soil concentration, ppm	N	Average value			CV, %		
			K_d	K_{oc}	K_{rs}^a	K_d	K_{oc}	K_{rs}^a
Bifenthrin	<0.5	16	3,570	237,000	18.1	47.6	29.4	25.2
Cyfluthrin	<0.5	4	1,370	124,000	9.56	21.5	35.3	40.0
Cypermethrin	0.5–1	3	14,200	310,000	24.1	77.7	60.9	67.4
Deltamethrin	0.5–1	19	2,300	704,000	21.8	72.0	73.5	44.0
Esfenvalerate								
Fenpropathrin	All Concentrations	15	454	42,500	2.79	113.3	74.8	80.9
Lambda-cyhalothrin	1–5	20	4,250	326,000	24.5	45.7	52.2	45.0
Permethrin	>5	16	2,230	277,000		25.1	55.4	
Tralomethrin	0.5–1	8	1,390	295,000	18.0	29.0	38.6	38.4

CV, coefficient of variation.

[a] K_{rs} is according to the specific surface method first proposed by Pionke and DeAngelis (1980) and later described by Wolt (1996) in a personal communication to the FIFRA Exposure Model Validation Task Force.

Table 3. Comparison of pyrethroid soil adsorption and desorption coefficients over all concentrations in experiments that measured both adsorption and desorption.

	Average value			
Chemical	K_d	K_{dd}	K_{oc}	K_{ocd}
Bifenthrin	3,570	9,070	237,000	622,000
Cyfluthrin	1,370	1,360	124,000	122,000
Cypermethrin	3,430	3,630	97,300	112,000
Deltamethrin	2,500	3,510	908,000	1,160,000
Esfenvalerate				
Fenpropathrin	453	378	41,700	40,000
Lambda-cyhalothrin	4,350	4,870	333,000	386,000
Permethrin	664	874	81,600	93,800
Tralomethrin	1,330	1,980	285,000	446,000

tion information. Data for esfenvalerate are absent from the tables because sorption experiments were not available for this chemical.

Values shown in Tables 2 and 3 were chosen by the selection criteria of experimental rating and sorbed soil concentrations. Whenever possible, average partition values from experiments with ratings greater than 4 and soil concentrations between 0.5 and 1 ppm were given preference. If there was no experiment with rating above 4, then the average value from the 0.5–1 ppm range was the next choice. In Table 2, the actual concentration range for each value is included, as are the number of values that comprised the adsorption average. Readers are referred to separate pyrethroid sections for additional details on the selection of sorption and desorption values.

Adsorption partition coefficients in Table 2 are presented in terms of total

Table 4. Freundlich adsorption data for pyrethroids.

	Average value			CV, %		
Chemical[a]	K_{df}	K_{ocf}	$1/n$	K_{df}	K_{ocf}	$1/n$
Bifenthrin						
Cyfluthrin						
Cypermethrin	8,920	281,000	0.999	50.1	60.4	7.3
Deltamethrin	6,240	2,840,000	1.13	84.5	113.0	16.0
Esfenvalerate						
Fenpropathrin	55.1	10,600	0.775	147.0	174.3	20.9
Lambda-cyhalothrin	5,490	4,180	0.962	177.4	91.5	11.7
Permethrin						
Tralomethrin	3,300	479,000	1.02	117.3	76.3	0.21

[a]Chemicals that are blank had no Freundlich data available.

soil surface area as well as soil organic carbon content. These coefficients, labeled "K_{rs}" in the table, are K_d divided by a measure of soil surface area estimated from organic carbon, sand, silt, and clay content of the soil used in a K_d measurement. The calculation of surface area was according to the methodology proposed by Pionke and DeAngelis (1980) and later communicated to the FIFRA Exposure Model Validation Task Force by Wolt (1996):

$$K_{rs} = \frac{0.01 \times K_d}{OC + (0.02 \times Clay) + (0.004 \times Silt) + (0.00005 \times Sand)} \quad (2)$$

where: K_d = adsorption partition coefficient, OC = % soil organic carbon content, $Clay$ = % clay, $Silt$ = % silt, and $Sand$ = % sand.

Sand, silt, and clay percentages were not available for the permethrin dataset in Table 2, and therefore K_{rs} for this material is not shown.

K_d values in Table 2 indicate all the pyrethroids are sorbed highly to soil; values are >1000 with the exception of fenpropathrin. This chemical has a K_d a factor of 10 lower than the rest, but this may result from experimental artifacts during measurement rather than pyrethroid sorption property. The fenpropathrin experiments were confounded by the breakdown of fenpropathrin to more polar and thus less highly adsorbed material during the equilibration portion of the sorption studies.

It is noted that, contrary to most findings, the expression of sorption as a function of soil organic carbon (K_{oc}) in Table 2 had little or no impact on reducing the variability of sorption from one soil to another. Variability as represented by coefficients of variation (CV) for average K_d and K_{oc} values is nearly the same with no clear trend; correcting sorption for soil surface area (K_{rs}) reduced variability slightly but perhaps not enough to be worthwhile when the greater need for data (sand, silt, and clay content in addition to soil organic carbon) is considered.

Table 3 summarizes and compares desorption to adsorption for all pyrethroids. So that adsorption could be compared to desorption directly, data were not filtered by the selection criteria described earlier but were filtered to exclude adsorption data not having corresponding desorption datasets. The adsorption and desorption partition coefficients are average values of data from all the concentrations for only experiments having measurements for both adsorption and desorption. Therefore, the adsorption averages presented in Table 3 are different from those cited in Table 2, which also included adsorption experiments lacking the corresponding desorption datasets. In general, desorption coefficients (K_{dd} and K_{ocd}) tend to be higher than adsorption coefficients (K_d and K_{oc}).

Nearly all pyrethroid sorption experiments were designed to evaluate concentration effects on sorption, and so Freundlich sorption data were included in this review whenever possible. These data are summarized in Table 4, which uses K_{df} to represent the Freundlich partition coefficient, K_{ocf} the Freundlich partition coefficient based on soil organic carbon content, and $1/n$ the Freundlich indicator of the concentration effect on sorption. If the value of $1/n$ is near 1.0, there is little or no impact of concentration on the sorption process. This is the case for $1/n$ in Table 4, and in retrospect, therefore, sorption data in this review could have been combined without regard to concentration because concentration had little apparent impact on sorption. As can be noted in the separate pyrethroid sections, differences in results between concentration-grouped and -ungrouped datasets are slight and probably not of great significance.

B. Abiotic and Biotic Kinetic Analyses

The sources of material for this review used a variety of kinetic procedures to estimate rates of degradation. To establish better uniformity in the kinetic treatment of the data, datasets from all abiotic and biotic degradation studies were reanalyzed with a standard procedure, assuming that rate of degradation was described by first-order kinetics. During the analyses it was noted that first order was not always obeyed, particularly for aerobic soil and soil photolysis datasets, since linear first-order plots of log-transformed data frequently displayed a curvilinear nature for data points instead of the expected straight-line relationship. These plots also frequently intercepted the y-axis at time $t = 0$ at values less than initial concentrations. Therefore, it was decided to plot datasets nonlinearly as well as linearly (nonlinear regression versus log-transformed linear regression) for all that had sufficient data. The curve-fitting program TableCurve 2D version 4.0 by Jandel Scientific Software (San Rafael, CA) provided the tool for the analyses and served as the basis for standardization via the program's curve-fitting routines for nonlinear and linear regression analyses.

Whenever possible, the results from nonlinear regression were selected for representation of a pyrethroid because, in general, nonlinear regression displayed better agreement over a larger portion of a degradation curve than did linear regression when first order was not obeyed. Typically, nonlinear fit was better for the first 80%–90% loss of parent and linear fit for the remaining 10%–20%.

For purposes of illustration, an example dataset is provided in Figs. 1 and 2. These plots were developed with TableCurve 2D and include TableCurve's equation number and equation that were selected to represent the nonlinear and linear first-order fits to a degradation dataset. In addition, the r-squared value indicating fit to data is given, along with the coefficients a and b, which are the values of the y-intercept at $t = 0$ and slope, respectively. The a coefficient is either C_0 directly (nonlinear) or the natural log of C_0 (linear) at time $t = 0$, as calculated by the regression; the b coefficient is the first-order rate constant, and from b the half-life (HL) was calculated by the equation:

$$HL = \frac{Ln2}{b} \tag{3}$$

All abiotic and biotic degradation datasets having three or more time intervals were analyzed with this procedure.

A few datasets consisted of only two sampling periods: time zero and some

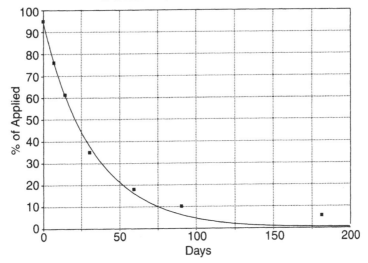

Degradation of Cyhalothrin in 18 Acres 10/31
Eqn 8098   y=aexp(-bx)
r²=0.99235842 DF Adj r²=0.98853763 FitStdErr=3.3510829 Fstat=649.31492
a=94.124443   b=0.029801614

Fig. 1. Example of TableCurve nonlinear first order plot of cyhalothrin degradation in soil identified as 18 acres 10/31. Squares are measured values of the cyhalothrin remaining at the specified time intervals.

later time interval. To include kinetics information from these datasets, it was assumed that first order applied, and a special form of the first-order equation integrated between the limits c_1 at t_1 and c_2 at t_2 was used to calculate first-order rate constants:

$$k = \frac{2.303}{t}\log\frac{c_1}{c_2} \qquad (4)$$

where: k = first-order rate constant in units of day^{-1}, t = the number of days between c_1 and c_2, and c_1, c_2 = concentrations of parent at times t_1 and t_2. In this review, Eq. 4 is referred to as the first-order point equation.

When several rate constants or half-life values were available, the data were condensed into a single value (expressed as half-life in the following tables) by calculation of the average of either the rate constant or the half-life for a given dataset. Averages for hydrolysis and photolysis datasets utilized the first-order rate constants to calculate the mean values, and these were then converted to corresponding half-life values, which are the values cited in the tables. However, for aerobic soil, anaerobic soil, and aerobic aquatic datasets, the averages were calculated directly from half-lives (not rate constants) of datasets, and these are the values given in the tables.

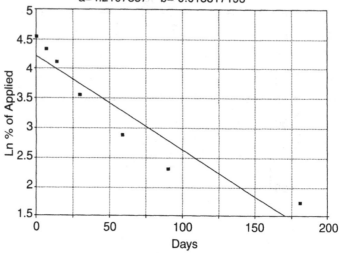

Fig. 2. Example of TableCurve linear first order plot of cyhalothrin degradation in soil identified as 18 acres 10/31. Dataset is same as in Fig. 1.

For aerobic soil, anaerobic soil, and aerobic aquatic datasets, a 90th percentile upper confidence limit was also calculated for each average from the following equations:

$$N = 1: \quad Avg + CL = Avg \times 3 \tag{5}$$

$$N > 1: \quad Avg + CL = Avg + \left(\frac{T_{90} \times STD}{\sqrt{N}} \right) \tag{6}$$

where: Avg = mean half-life value of a dataset assuming normal distribution, $CL = 90^{th}$ percentile upper confidence limit, T_{90} = Student's 90th percent t-value (one-tailed) for a given value of N, N = number of values in a dataset, and STD = standard deviation of the mean of N values.

C. Abiotic Chemical Properties

1. Hydrolysis. Table 5 presents the hydrolysis rates of pyrethroids in water at acidic, neutral, and alkaline pH. As indicated earlier, the values in the table were developed by averaging nonlinear first-order rate constants for a given pH and then calculating the half-life in Table 5 from the average rate.

With the exception of tralomethrin, whose data were compromised during analysis by incomplete transfer of tralomethrin from buffer to organic solvent

Table 5. Hydrolysis properties of pyrethroids at 25 °C.

Chemical	Hydrolysis half-life, d		
	pH 5	pH 7	pH 9
Bifenthrin	Stable	Stable	Stable
Cyfluthrin	Stable	183	1.84
Cypermethrin	619	274	1.90
Deltamethrin	Stable	Stable	2.15
Esfenvalerate	Stable	Stable	Stable
Fenpropathrin	1670	555	14.4
Lambda-cyhalothrin	Stable	Stable	8.66
Permethrin	Stable	Stable	242
Tralomethrin	167	15.5	38.9

as well as conversion of tralomethrin to deltamethrin on day 0, pyrethroids are degraded slowly at acidic and neutral pH. Hydrolysis is more rapid in alkaline water for all except bifenthrin and esfenvalerate, which are stable to hydrolysis at all three pH values. The bifenthrin hydrolysis data have a certain degree of uncertainty due to the experimental design used to measure hydrolysis. Only one experiment was performed (Herbst 1983c), and it was conducted at concentrations far in excess of bifenthrin water solubility (crystal formation observed), casting some uncertainty on the true hydrolytic stability of bifenthrin.

With the exceptions just described, pyrethroid hydrolysis studies generally had no issues that impacted hydrolysis results significantly. Except for bifenthrin and perhaps tralomethrin, the data in Table 5 can generally be assumed to reflect the hydrolytic stability of pyrethroids in water. For further details, the reader is referred to individual pyrethroid sections.

2. Photolysis. Pyrethroid water photolysis experiments did not always utilize the same source of light for irradiation, and thus experiments were separated on the basis of outdoor experiments under natural sunlight versus laboratory under artificial light. An additional factor that may have had impact on the results was the use of emulsifier or cosolvents other than acetonitrile in the preparation of solutions for irradiation. To account for this impact, experiments were categorized according to "buffer medium only (includes acetonitrile)" or medium having cosolvent (not acetonitrile) or emulsifier in addition to the sunlight/artificial light classification.

Soil photolysis studies have the same issue of light source. In addition, a few studies were impacted by the method of pyrethroid application; some experimenters drenched the chemical into the soil with large volumes of organic solvent or mixed the chemical into the soil after application. This method could render a portion of chemical unavailable to sunlight and cause datasets to deviate from first-order kinetics because of the shielding effect of the soil to irradiation.

A final issue in soil photolysis studies was the soil moisture status during irradiation. Most studies were conducted under air-dry soil conditions; a few were carried out at moisture values near field capacity. Results were not always the same. The possibility is raised that soil moisture may influence photolysis on soil surfaces and that rates might be slower when the surface of the soil is wet, which seldom is the case during periods of sunshine.

Table 6 summarizes the photolysis results for water and soil studies. All the studies included dark control samples, and therefore photolysis half-lives cited in the table have been corrected for dark control reaction rate. Data in the table also were adjusted for continuous irradiation with artificial light by comparison to the intensity of natural sunlight and then making appropriate adjustments to the kinetics, or by assuming 12-hr light and dark periods and reducing rate constants by a factor of 2.

Photolysis in water yielded half-lives from a low of 0.7 d for cyfluthrin to highs of 408 d for bifenthrin and 603 d for fenpropathrin. With respect to bifenthrin, its photolysis was investigated with natural sunlight as well as artificial light. The 408-d value cited in Table 6 is from the natural sunlight experiment, but the artificial light study had a half-life of 25 d. These results are contradictory, and it was decided to select the more conservative result from natural sunlight to represent bifenthrin. The possibility of more rapid photolysis cannot be totally excluded, however, and if such is the case, the bifenthrin rate may be more in line with its soil photolysis rate and with those of the other pyrethroids.

Table 6. Photolysis of pyrethroids in water or on soil surfaces.

Chemical	Photolysis half-life, d		
		Soil	
	Water	Air dry	Wet
Bifenthrin	408	96.9	
Cyfluthrin	0.673	5.02	
Cypermethrin	30.1		165
Deltamethrin	55.5		34.7
Esfenvalerate	17.2	10.0[a]	No photolysis
Fenpropathrin	603	4.47[b]	No photolysis
Lambda-cyhalothrin	24.5	53.7	
Permethrin	110	104	
Tralomethrin	2.47	3.87	

[a]Mean of two values from air-dry soil and the single wet soil value from Table 81, assuming $k = 0$.

[b]Mean of three values from air-dry soil and the single wet soil value from Table 94, assuming the wet soil $k = 0$.

Results of the soil photolysis studies in Table 6 indicate that pyrethroids can be susceptible to photolysis when on soil surfaces, suggesting this may be a significant dissipation mechanism for any pyrethroid that reaches exposed soil surfaces. There also is the suggestion that photolysis on dry soil surfaces may be faster than on wet surfaces. Half-lives overall tend to be less than 55 d except for bifenthrin, cypermethrin, and permethrin. In the case of cypermethrin, it has been studied only in the presence of wet soil, and it is possible that a more rapid rate of photolysis would have been observed with soil in an air-dry state, because two pyrethroids, esfenvalerate and fenpropathrin, had rapid photolysis on air-dry soil whereas no photolysis was observed on wet soil.

D. Biotic Chemical Properties

1. Aerobic Soil Degradation.   Occasionally pyrethroid aerobic soil degradation experiments were confounded by the application of pyrethroid in volumes of organic solvent large enough to be potentially inhibitory to soil microflora. In addition, some experiments used pyrethroid treatment rates in excess (ppm range) of that which would normally be applied for pest control. Either of these factors could impact the degradation kinetics so that the data might not reflect what actually takes place in the field under conditions of normal application. These potentially confounding factors were considered during assessment of experimental quality through the use of a rating system that assigned a value of 1 to 10 (10 being the highest quality) to each experiment. In addition, pyrethroid degradation results were organized by treatment application categories of <0.5 ppm, 0.5–1.5 ppm, >1.5 to 10 ppm, and >10 ppm to account for the possible impact of treatment rate.

Experiments had been conducted at temperatures ranging from 10° to 40 °C. Because of this wide range, the data were filtered to include only experiments performed between 15° and 30 °C; typically, 20°–28 °C. This range was selected because it is the one commonly encountered in soils during active plant growth periods, and it allowed the greatest use of the degradation data for purposes of comparison while at the same time minimizing the impact of temperature. Soil moisture levels, another potential influence on soil degradation kinetics, were not a significant factor because typically studies were carried out at optimum or near-optimum moisture content close to soil field capacity. Thus, datasets were not categorized by moisture content and can be assumed to have been performed at optimal or near-optimal soil moisture conditions.

Table 7 is a summation of the aerobic half-life data abstracted from the individual pyrethroid sections. Values in the table are averages of half-life datasets selected only from experiments having a confidence rating >4 so that solvent and concentration effects are minimized. Also included are the number of datapoints in each average and the average upper confidence limits at the 90th percentile level, as calculated by Eq. 5 or 6. Half-lives range from a low of 3.25 d for tralomethrin to a high of 96.3 d for bifenthrin.

Table 7. Degradation of pyrethroids in aerobic soil.

Chemical	N	Half-life, d	
		Mean	Mean + CL
Bifenthrin	3	96.3	125
Cyfluthrin	2	11.5	28.0
Cypermethrin	7	27.6	39.2
Deltamethrin	9	24.2	27.4
Esfenvalerate	9	38.6	45.2
Fenpropathrin	4	22.3	31.5
Lambda-cyhalothrin	5	42.6	58.9
Permethrin	8	39.5	58.7
Tralomethrin	1	3.25	9.75

CL, confidence limit.

2. Anaerobic Degradation. This review separates the anaerobic degradation of pyrethroids into two categories on the basis of experimental design. The first category is based on the anaerobic soil design, which subjects the pyrethroid to an initial period of aerobic degradation (approximately 1 half-life or 30 d) at moisture levels near field capacity, and then continues on to anaerobicity through waterlogging, exchange of incubation container headspace to nitrogen gas, and subsequent incubation anaerobically for another 30–60 d. The second category is the anaerobic aquatic design in which the pyrethroid is applied to waterlogged soil–sediment systems made anaerobic by prior incubation under nitrogen, or applied to systems at the time of waterlogging and purging of the incubation container headspace with nitrogen gas. Anaerobic incubation is continued with periodic sampling for periods up to 1 yr.

Both types of experiments followed a data evaluation pathway similar to the aerobic soil studies. High application rate was an issue for some experiments, but solvent effects were of lesser importance due to the added benefit of additional water that served to dilute any cosolvent added initially. Experiments were assigned experimental ratings in the same manner as aerobic studies, and data from experiments with experimental ratings >4 were selected for Tables 8 and 9, which compile the anaerobic results for the pyrethroids. Data were collected from the individual pyrethroid sections, and for more details the reader is referred to those sections.

Table 8 presents the results from anaerobic soil experimental design. Mean half-lives and 90th percent confidence limits are available for all pyrethroids except lambda-cyhalothrin. Table 9 gives the mean half-lives, the number of values in each mean, and the 90th percent confidence of the mean for the datasets from anaerobic aquatic experimental designs. The absence of data means that no experiments were conducted with an anaerobic aquatic design. In general,

Table 8. Anaerobic degradation of pyrethroids from studies using an anaerobic soil experimental design.

Chemical	N	Half-life, d	
		Mean	Mean + CL
Bifenthrin	1	425	1280
Cyfluthrin	1	33.6	101
Cypermethrin	1	55.0	165
Deltamethrin	2	28.9	36.8
Esfenvalerate	1	90.4	271
Fenpropathrin	3	276	386
Lambda-cyhalothrin			
Permethrin	1	197	591
Tralomethrin	1	5.00	15.0

pyrethroids continue to degrade in anaerobic environments with rates of degradation similar to those in aerobic systems.

3. Aerobic Aquatic Degradation.   Available pyrethroid degradation rates in aerobic aquatic systems are summarized in Table 10, which presents mean half-lives of datasets for each pyrethroid, the number of values in each mean, and the corresponding 90th percent confidence limit of the mean calculated from Eq. 5 or 6.

Aerobic aquatic experiments did not have the issues of high application rate and use of cosolvents that were present in aerobic soil and anaerobic degradation experiments. Not all experiments were run at the same incubation temperature,

Table 9. Anaerobic degradation of pyrethroids from studies using an anaerobic aquatic experimental design.

Chemical	N	Half-life, d	
		Mean	Mean + CL
Bifenthrin			
Cyfluthrin			
Cypermethrin	1	7.4	22.2
Deltamethrin			
Esfenvalerate			
Fenpropathrin	2	73.6	90.8
Lambda-cyhalothrin			
Permethrin			
Tralomethrin			

Table 10. Degradation of pyrethroids in aerobic aquatic systems.

Chemical	Temperature, °C	N	Half-life, d	
			Mean	Mean + CL
Bifenthrin				
Cyfluthrin				
Cypermethrin	16–25	7	7.44	10.6
Deltamethrin	20	2	79.5	189
Esfenvalerate	10	2	72.3	93.5
Fenpropathrin				
Lambda-cyhalothrin	20	4	21.9	28.2
Permethrin				
Tralomethrin				

as can be noted in Table 10, but because data were not extensive, the choice was made to pool data for a given pyrethroid regardless of the incubation temperature. Degradation rates in aquatic systems varied from a half-life of 7.4 d (cypermethrin) to 79.5 d (deltamethrin). These rates are similar to those noted for degradation in aerobic soils. No information was available for bifenthrin, cyfluthrin, fenpropathrin, permethrin, or tralomethrin.

The summary and comparative review of the physical and chemical properties of the pyrethroids are now concluded. The sections that follow present information for each individual pyrethroid in greater detail and serve as the bases for the data presented here.

III. Bifenthrin

Bifenthrin [Chemical Abstracts (C.A.) = (2-methyl[1,1′-biphenyl]-3-yl)methyl 3-(2-chloro-3,3,3-trifluoro-1-propenyl)-2,2-dimethylcyclopropanecarboxylate and IUPAC = 2-methylbiphenyl-3-ylmethyl(Z)-(1RS,3RS)-3-(2-chloro-3,3,3-trifluoroprop-1-enyl)-2,2-dimethylcyclopropanecarboxylate] consists of an isomeric mixture of 97% cis isomer and 3% trans form. Its molecular weight is 422.9. FMC Corporation developed the material under the code name FMC 54800; its CAS RN is 82657–04-3, and Official code = OMS 3024 (Tomlin 1994).

A. Physical Properties

1. Vapor Pressure.   No reports measuring the vapor pressure of bifenthrin were reviewed; the value cited in Table 11 was taken from the citation in *The Pesticide Manual*, 10th Ed. (Tomlin 1994).

2. Water Solubility.   Herbst (1983a) measured the water solubility of bifenthrin with a column saturation method that generated saturated solutions of material.

Table 11. Bifenthrin vapor pressure.

Temperature, °C	Best value	VP, mm Hg	Source
25	No	$1.80E-07$[a]	Tomlin (1994)

[a]Representative value.

The solutions were passed directly through collector columns composed of Brownlee Labs 18-GU RP 18 guard columns, which were then analyzed for bifenthrin content. The value cited in Table 12 is the average of six measurements.

3. Henry's Law Constant. Henry's law constant was calculated from Eq. 1, assuming molecular weight $= 422.9$, vapor pressure $= 1.8 \times 10^{-7}$ mm Hg, and water solubility $= 1.4 \times 10^{-5}$ ppm. K_h from this calculation is 7.2×10^{-3} atm m^3 mol^{-1}.

4. Octanol–Water Partition Coefficient. Because of the very large difference in concentration of pyrethroid between the octanol and water phases, experimenters had to pay close attention to the minimization of octanol contamination of the water phase from suspended droplets or emulsion formation. Any octanol left in the water phase and included in the assay for pyrethroid could cause potentially large underestimations of K_{ow}. Therefore, from descriptions of methodology there were judgments on the levels of confidence reflected in the K_{ow} values for this and all other pyrethroids; this is reflected by the "Best values" column shown in Table 13.

Bifenthrin partitioning between octanol and water was studied by Herbst (1983b), who equilibrated bifenthrin in octanol with water for 30 min with stirring and then attempted to analyze both phases for concentration of bifenthrin. Analy-

Table 12. Water solubility of bifenthrin.

Temperature, °C	Column saturation	WS, ppm	Reference
22	Yes	$1.40E-05$[a]	Herbst (1983a)

[a]Representative value.

Table 13. Bifenthrin octanol–water partition coefficient.

Best values	K_{ow}	Log P	CLog P	Reference
	$3.00E+06$	6.4[a]	7.2[b]	Herbst (1983b)

[a]Representative value.
[b]Calculated from molecular structure by Briggs (1999).

Table 14. Bifenthrin bioconcentration factor based on fish whole-body analysis.

Type of experiment	Best values	BCF	Reference
Flow-through	Yes	6090[a]	Suprenant (1986)

[a]Representative value.

sis of the water phase failed as a result of concentrations lower than method sensitivity (0.003 ppm), and K_{ow} was estimated indirectly using established correlations between K_{ow} and water solubility. K_{ow} as Log P from this procedure is the value cited in Table 13. Log P calculated from bifenthrin molecular structure (CLog P) by Briggs (1999) is also presented in Table 13 for purposes of comparison, and it is noted that CLog P is larger than Log P. Because of the resultant uncertainty in K_{ow}, the "Best values" column in the table is left blank.

5. Bioconcentration Factor.   Suprenant (1986) exposed bluegill sunfish continuously to bifenthrin in flow-through aquaria up to 42 d (Table 14). Measurements of total ^{14}C in fish and water were made throughout the exposure period, and BCF listed in Table 14 was calculated from total ^{14}C concentrations in fish and water on day 42 of the study, assuming all ^{14}C was bifenthrin.

6. Soil Sorption.   Sorption of bifenthrin to soil was measured by Froelich (1983), who prepared several solutions of bifenthrin in calcium chloride containing 2% acetonitrile as cosolvent and equilibrated them with soil in glass flasks for 24 hr. After centrifugation the supernatant was assayed for total ^{14}C, and soil-sorbed material was calculated from the difference in water concentration before and after equilibration. K_d was then calculated from these measurements, assuming all ^{14}C in water and soils was bifenthrin. Desorption was measured by removal of supernatant from adsorption samples, replacement with fresh calcium chloride, and reequilibration for an additional 24 hr.

Table 15 cites the adsorption data and Table 16 the desorption data produced by this experiment. The inclusion of cosolvent in the experimental design and no identification of ^{14}C may have caused underestimation of sorption coefficients and are the reasons for the lower experimental ratings listed in both tables. With regard to Freundlich sorption data, no information on the adsorption at more than one concentration for a given soil was available, and Freundlich coefficients could not be calculated.

B. Abiotic Chemical Properties

1. Hydrolysis.   Herbst (1983c) added bifenthrin in acetonitrile (7.7% cosolvent) to buffer solutions at concentrations above its solubility (0.5 or 5 ppm). Crystal formation was observed during subsequent incubation, and assay of entire samples showed little or no change in total bifenthrin throughout the course of the

Table 15. Bifenthrin soil adsorption partition coefficients.

Expt rating	Soil conc, ppm	Soil texture	K_d	K_{oc}	Reference
3	0.2	FS	882	116,000	Froelich (1983)
3	0.2	FS	1,040	137,000	Froelich (1983)
3	0.2	FS	1,110	146,000	Froelich (1983)
3	0.2	FS	941	124,000	Froelich (1983)
3	0.4	CL	3,550	265,000	Froelich (1983)
3	0.4	SL	3,540	203,000	Froelich (1983)
3	0.4	SL	4,640	267,000	Froelich (1983)
3	0.4	SL	4,620	266,000	Froelich (1983)
3	0.4	SiL	5,230	291,000	Froelich (1983)
3	0.4	SiL	5,230	291,000	Froelich (1983)
3	0.4	SL	3,840	221,000	Froelich (1983)
3	0.4	SiL	5,910	328,000	Froelich (1983)
3	0.4	CL	3,860	288,000	Froelich (1983)
3	0.4	CL	3,780	282,000	Froelich (1983)
3	0.4	CL	3,560	266,000	Froelich (1983)
3	0.4	SiL	5,340	297,000	Froelich (1983)
All ratings	Avg, <0.5		*3,570*[a]	*237,000*[a]	
	CV, %		47.6	29.4	
	Avg, 0.5 to <1				
	CV, %				
	Avg, 1–5				
	CV, %				
	Avg, >5				
	CV, %				
	Avg, All Conc		3,570	237,000	
	CV, %		47.6	29.4	
>4	Avg, <0.5				
	CV, %				
	Avg, 0.5 to <1				
	CV, %				
	Avg, 1–5				
	CV, %				
	Avg, >5				
	CV, %				
	Avg, All Conc				
	CV, %				

Expt: experiment; Conc: concentration; Avg: average; FS: fine sandy loam; CL: clay loam; SL: sandy loam; SiL: silt loam.
[a]Representative value.

Table 16. Bifenthrin soil desorption partition coefficients.

Expt rating	Soil concentration, ppm	Soil texture	K_{dd}	K_{ocd}	Reference
3	0.2	FS	3,430	451,000	Froelich (1983)
3	0.2	FS	2,840	374,000	Froelich (1983)
3	0.2	FS	3,670	483,000	Froelich (1983)
3	0.2	FS	3,430	451,000	Froelich (1983)
3	0.4	CL	9,550	713,000	Froelich (1983)
3	0.4	SL	9,920	570,000	Froelich (1983)
3	0.4	SL	12,300	707,000	Froelich (1983)
3	0.4	SL	12,200	701,000	Froelich (1983)
3	0.4	SiL	11,100	617,000	Froelich (1983)
3	0.4	SiL	12,100	672,000	Froelich (1983)
3	0.4	SL	9,770	561,000	Froelich (1983)
3	0.4	SiL	10,900	606,000	Froelich (1983)
3	0.4	CL	9,550	713,000	Froelich (1983)
3	0.4	CL	11,900	888,000	Froelich (1983)
3	0.4	CL	10,100	754,000	Froelich (1983)
3	0.4	SiL	12,400	689,000	Froelich (1983)
All ratings	Avg, <0.5		9,070	622,000	
	CV, %		39.3	21.6	
	Avg, 0.5 to <1				
	CV, %				
	Avg, 1–5				
	CV, %				
	Avg, >5				
	CV, %				
	Avg, All Conc		9,070	622,000	
	CV, %		39.3	21.6	
>4	Avg, <0.5				
	CV, %				
	Avg, 0.5 to <1				
	CV, %				
	Avg, 1–5				
	CV, %				
	Avg, >5				
	CV, %				
	Avg, All Conc				
	CV, %				

Table 17. Hydrolysis of bifenthrin in buffered water.

Temperature, °C	pH	Nonlinear k[a]	Half-life, d[a]	Linear k[a]	Half-life, d[a]	Reference
25	5	$0.00E + 00$[b]	0	$0.00E + 00$	0	Herbst (1983c)
25	7	$0.00E + 00$[b]	0	$0.00E + 00$	0	Herbst (1983c)
25	9	$0.00E + 00$[b]	0	$0.00E + 00$	0	Herbst (1983c)

[a]An entry of $0.00E + 00$ means k was measured but bifenthrin is stable. A blank k or half-life value means no measurement.
[b]Representative value.

study. There was no evidence of hydrolysis product formation at any value of pH. The study provides no estimate of hydrolysis rate; the rate is presumed to be "stable" and is cited as such in Table 17.

2. Photolysis in Water.   The photolysis of bifenthrin in water irradiated by natural sunlight or by continuous irradiation with a GE sunlamp was reported by Wu et al. (1986a). In both cases, bifenthrin in acetonitrile was added to sterile buffer (30% acetonitrile) in borosilicate ampules that were then sealed and exposed to natural sunlight up to 30 d and up to 14 d under the GE sunlamp. Natural light was late summer New Jersey sunshine. Data are summarized in Table 18 for both studies; the data are corrected for dark reaction kinetics from dark controls included in the study. The data from artificial light are also adjusted for continuous exposure by assuming 12-hr light and dark periods.

3. Photolysis on Soil.   Wu et al. (1986b) measured the photolysis of bifenthrin on soil surfaces (Table 19). Bifenthrin in 2 mL of methylene chloride was applied to the surface of sterilized air-dry soil in petri dishes sealed with pyrex glass plates. Half the dishes were covered with aluminum foil to provide dark controls, and the remainder were exposed to New Jersey summer sunlight for up to 30 d. Data from the first-order nonlinear and linear analyses of the study

Table 18. Photolysis of bifenthrin in water.

Natural sun	Buffer medium	Nonlinear k	Half-life, d	Linear k	Half-life, d	Reference
Yes	Yes	$1.70E - 03$[a]	408	$1.68E - 03$	413	Wu et al. (1986a)
No	Yes	$2.75E - 02$	25.2	$2.91E - 02$	23.8	Wu et al. (1986a)

[a]Representative value.

Table 19. Photolysis of bifenthrin on soil.

Natural sun	Moisture state	Nonlinear		Linear		Reference
		k	Half-life, d	k	Half-life, d	
Yes	Air dry	$7.15E - 03$[a]	96.9	$6.96E - 03$	99.6	Wu et al. (1986b)

[a]Representative value.

dataset are presented in Table 19; these data have been corrected for dark reaction.

C. Biotic Chemical Properties

1. Aerobic Soil Degradation.   Two studies were reviewed that reported on the degradation of bifenthrin in aerobic soils (Table 20). Bixler et al. (1983) added bifenthrin in 150 μL ethanol to 50 g soil in biometer flasks. After solvent evaporation and mixing, water was added to 65% field capacity, and samples were incubated at 25 °C in the dark for up to 61 d. Reynolds (1984) added the chemical at a higher application rate in 204 μL ethanol to 50 g soil in biometer flasks. Samples were then incubated in a similar fashion for up to 180 d. Results from nonlinear and linear analyses of datasets from both studies are summarized in Table 20; the study by Reynolds is assigned a lesser experimental rating on the basis of higher treatment rate and greater volume of solvent.

2. Anaerobic Soil Degradation.   One report was reviewed that utilized an anaerobic soil experimental design to study the degradation of bifenthrin in anaerobic systems (Table 21). Reynolds (1986) added bifenthrin in 180–320 μL ethanol to 50 g soil in biometer flasks, which were incubated aerobically in the dark for 29 d. Samples were then made anaerobic by flooding with water and amendment with ground alfalfa to aid in establishment of anaerobicity. Incubations continued for up to 61 d. The study experimental design was assigned an experimental rating of 4 on the basis of high application rate applied in a fairly large volume of ethanol. Samples were analyzed only on days 31 and 61 d after flooding, and so the first-order point equation (Eq. 3) was used to estimate the linear first-order rate constant in Table 13 because data were insufficient for regression analysis.

3. Aerobic Aquatic Degradation.   No information was available for the degradation of bifenthrin in aerobic aquatic systems.

4. Anaerobic Aquatic Degradation.   No studies were reviewed that utilized an anaerobic aquatic experimental design.

Table 20. Bifenthrin degradation in aerobic soil.

Expt rating	Temp, °C	Soil conc, ppm	Soil texture	Nonlinear		Linear		Reference
				k	Half-life, d	k	Half-life, d	
6	25	1	SL	8.04E − 03	86.2	8.04E − 03	86.2	Bixler et al. (1983)
6	25	1	SiL	5.48E − 03	126	5.32E − 03	130	Bixler et al. (1983)
6	25	1	SiCL	9.04E − 03	76.7	8.93E − 03	77.6	Bixler et al. (1983)
3	25	3.2	SL	7.55E − 03	91.8	5.26E − 03	132	Reynolds (1984)
3	25	3.2	SiL	2.75E − 03	252	2.78E − 03	249	Reynolds (1984)
3	25	3.2	SiCL	5.13E − 03	135	5.37E − 03	129	Reynolds (1984)
All ratings		Avg, <0.5 CV, % Avg + CL						
		Avg, 0.5–1.5			96.3		97.9	
		CV, %			27.2		28.7	
		Avg + CL			125		129	
		Avg, 1.5–10			160		170	
		CV, %			51.9		40.3	
		Avg + CL			250		245	
		Avg, >10 CV, % Avg + CL						
		Avg, All Conc			128		134	
		CV, %			50.8		45.7	
		Avg + CL			167		171	
>4		Avg, All Conc			96.3[a]		97.9	
		CV, %			27.2		28.7	
		Avg + CL			125[a]		129	

Temp: temperature.
[a]Representative value.

IV. Cyfluthrin

Cyfluthrin is a diastereoisomeric mixture of four pairs of enantiomers known as cyfluthrin1 (23%–26%), cyfluthrin2 (16%–19%), cyfluthrin3 (33%–36%), and cyfluthrin4 (22%–25%). [IUPAC = (*RS*)-α-cyano-4-fluoro-3-phenoxybenzyl (1*RS*, 3*RS*;1*RS*,3*SR*)-3-(2,2-dichlorovinyl)-2,2-dimethylcyclopropanecarboxylate]. It has sometimes been referred to as Baythroid®, and its development code is BAY FCR 1272. CAS RN is 68359-37-5; its molecular weight is 434.3 (Tomlin 1994).

Table 21. Bifenthrin anaerobic degradation generated by anaerobic soil experimental design.

Expt rating	Temp, °C	Soil conc, ppm	Soil texture	Nonlinear		Linear		Reference
				k	Half-life, d	k	Half-life, d	
4	25	3	SL			1.63E – 03	425	Reynolds (1986)
All ratings		Avg, <1.5 CV, % Avg + CL						
		Avg, >1.5 CV, % Avg + CL					425[a] 1280[a]	
		Avg, All Conc CV, % Avg + CL						
>4		Avg, All Conc CV, % Avg + CL						

[a]Representative value.

A. Physical Properties

1. Vapor Pressure. Vapor pressure of cyfluthrin has been measured by a static method using the vapor pressure balance procedure according to OECD Guidelines No. 104. The first series of measurements were made by Sewekow (1981) with cyfluthrin1, cyfluthrin2, cyfluthrin3, and cyfluthrin4 at elevated temperatures; results were then extrapolated to 20° and 25 °C. The second series (Talbot and Mosier 1987) utilized a synthetic mixture of the four enantiomer pairs in ratios approximating the same proportions in technical cyfluthrin. Temperature ranged from 20° to 40 °C, and results for 25 °C were interpolated from the corresponding temperature plot. Results for both series are listed in Table 22, and collectively it is assumed that the respective mean values at 20° and 25 °C represent cyfluthrin vapor pressure.

2. Water Solubility. Cyfluthrin water solubility has been measured on three separate occasions, all at 20 °C, by the column saturation procedure according to OECD Guidelines No. 105 (Krohn 1983a,b, 1987a,b, 1988). The first and second studies in 1983 and 1987 measured the separate solubilities of the cyfluthrin diastereoisomers cyfluthrin1, cyfluthrin2, cyfluthrin3, and cyfluthrin4; the third study measured the solubilities of cyfluthrin2 and cyfluthrin4. Results for

Table 22. Vapor pressure of cyfluthrin.

Temperature, °C	VP, mm Hg	Source
25	4.90E – 08	Talbot and Mosier (1987)
20	7.30E – 09	Sewekow (1981)
25	1.60E – 08	Sewekow (1981)
20	1.10E – 10	Sewekow (1981)
25	2.60E – 09	Sewekow (1981)
20	1.60E – 10	Sewekow (1981)
25	3.60E – 09	Sewekow (1981)
20	6.50E – 10	Sewekow (1981)
25	1.50E – 09	Sewekow (1981)
Avg, 20	2.1E – 09	
CV, %	170.6	
Avg, 25	*1.5E – 08*[a]	
CV, %	138.5	

[a]Representative value.

all measurements are presented in Table 23, and, because values are not distinctly different and the combination of enantiomers in cyfluthrin is roughly equivalent, it is assumed that their average value represents the water solubility of cyfluthrin.

3. Henry's Law Constant. Henry's law constant (K_h) for cyfluthrin is calculated by Eq. 1, assigning a 25 °C value of 1.5×10^{-8} mm Hg to vapor pressure. Water solubility at 25 °C was not available, and thus the value of 2.3×10^{-3} ppm at 20 °C from Table 23 was assigned to water solubility. The molecular weight of cyfluthrin is 434.3, and with these values as input, calculation of K_h by Eq. 1 provides the value of 3.7×10^{-6} atm m^3 mol^{-1}.

4. Octanol–Water Partition Coefficient. Cyfluthrin octanol–water partition coefficients (K_{ow}) have been measured by Krohn (1987a) for each of the diastereoisomers with a procedure that depends on equilibration of cyfluthrin between octanol and water after phases are shaken together in a cylinder. This method follows the OECD Guidelines method No. 107. Results from these measurements are summarized in Table 24 for each of the diastereoisomers, and the value for cyfluthrin is assumed to be reflected by the mean of the four isomer values. Although a shake procedure was used to equilibrate the water and octanol phases (the most susceptible to emulsion formation), Krohn paid close attention to the minimization of octanol contamination of the water phase from suspended droplets or emulsion formation. As stated earlier in the bifenthrin section, any octanol left in the water phase and included in the assay for cyfluthrin could cause potentially large underestimations of K_{ow}. Krohn's effort to solve this problem is the reason for the "Yes" values assigned to the "Best

Table 23. Cyfluthrin water solubility.

Temperature, °C	Column saturation	WS, ppm	Reference
20	Yes	2.50E – 03	Krohn (1987b)
20	Yes	2.20E – 03	Krohn (1987b)
20	Yes	2.10E – 03	Krohn (1987b)
20	Yes	1.90E – 03	Krohn (1987b)
20	Yes	3.20E – 03	Krohn (1987b)
20	Yes	2.20E – 03	Krohn (1987b)
20	Yes	4.30E – 03	Krohn (1987b)
20	Yes	2.90E – 03	Krohn (1987b)
20	Yes	2.06E – 03	Krohn (1988)
20	Yes	1.22E – 03	Krohn (1988)
20	Yes	1.90E – 03	Krohn (1983b)
20	Yes	2.30E – 03	Krohn (1983b)
20	Yes	1.80E – 03	Krohn (1983b)
20	Yes	1.90E – 03	Krohn (1983b)
Avg		*2.32E – 03*[a]	
CV, %		32.1	

[a]Representative value.

values" column in Table 24. CLog P calculated from the cyfluthrin molecular structure by Briggs (1999), provided in the table for comparison, is higher but similar to the actual measured value.

5. Bioconcentration Factor. Bioconcentration (Table 25) was measured in bluegill sunfish exposed to a continuous flow of water containing ^{14}C-cyfluthrin for periods up to 28 d (Carlisle and Roney 1984). Results were reported on the basis of total ^{14}C in whole-body fish. Radioactivity in the fish was characterized by partitioning against polar (acetonitrile) and nonpolar (hexane) solvents. Total

Table 24. Cyfluthrin octanol–water partition coefficient.

Best values	K_{ow}	Log P	CLog P	Reference
Yes	1.00E + 06	6.00		Krohn (1987a)
Yes	8.70E + 05	5.94		Krohn (1987a)
Yes	1.10E + 06	6.04		Krohn (1987a)
Yes	8.20 + 05	5.91		Krohn (1987a)
Avg	*9.48E + 05*[a]	*5.97*[a]	6.4[b]	
CV, %	13.4	1.0		

[a]Representative values.
[b]Calculated from molecular structure by Briggs (1999).

Table 25. Cyfluthrin bioconcentration factor based on fish whole-body analysis.

Type of experiment	Best values	BCF	Reference
Continuous flow	Yes	719[a]	Carlisle and Roney (1984)

BCF, bioconcentration factor.
[a]Value selected for model input.

residues were sufficiently low so that the researchers deemed it unnecessary to perform further characterization. The BCF value cited in Table 25 is the average of fish:water ^{14}C ratios calculated from days 10 through 28, which did not differ statistically.

6. Soil Sorption.   The sorption of cyfluthrin to soil was studied by Gronberg in 1987 and again by Burhenne in 1996. Gronberg's study did not use radioactive material, the presentation of experimental details was sketchy, and no balance of material was provided. It was considered not as good a study as the second one, which was an excellent design for pyrethroid soil sorption studies. The second study utilized ^{14}C-cyfluthrin, added $HgCl_2$ as a bactericide to minimize degradation of parent, and carried out identification analyses of both water and soil phases for all samples. It thus allowed direct calculation of partition coefficients without having to assume that all ^{14}C in the water phase was 100% parent and that the loss of ^{14}C from the water phase equated to parent material sorbed to soil.

A minor issue with this study was the use of 1% isopropanol cosolvent in the test solution to enhance cyfluthrin water solubility for purposes of overcoming analytical difficulties caused by the extremely low water solubilities. The presence of the cosolvent may have resulted in an underestimation of cyfluthrin sorption properties because of a possible impact on test solution polarity. Adsorption and desorption data for cyfluthrin are presented in Tables 26 and 27. Freundlich soil sorption data were not available for cyfluthrin.

B. Abiotic Chemical Properties

1. Hydrolysis.   Sandie (1983) examined the hydrolysis of cyfluthrin at 25 °C and pH values of 5, 7, and 9. Kinetic data from first-order nonlinear and linear plots of the data are presented in Table 28.

2. Photolysis in Water.   The photolysis of cyfluthrin in water has been characterized by Puhl et al. (1983), by Gronberg (1984), and again in 1990 by Hellpointner (Table 29). The Hellpointner study was designed to measure quantum yield according to the methodology outlined by the ECETOC Task Force on Photodegradation, and the kinetic information presented in Table 29 has not been adjusted for sunlight simulation or light/dark periods. When these adjust-

Table 26. Cyfluthrin soil adsorption partition coefficients.

Expt rating	Soil conc, ppm	Soil texture	K_d	K_{oc}	Reference
3	0.007	SL	633	50,200	Gronberg (1987)
3	0.007	SL	735	58,300	Gronberg (1987)
3	0.007	SL	1,060	84,400	Gronberg (1987)
5	0.1	CL	1,790	73,500	Burhenne (1996)
5	0.1	LS	1,320	118,000	Burhenne (1996)
5	0.1	LS	1,240	180,000	Burhenne (1996)
5	0.1	SiL	1,120	124,000	Burhenne (1996)
All ratings	Avg, <0.5		1,130	98,300	
	CV, %		34.2	46.3	
	Avg, 0.5 to <1				
	CV, %				
	Avg, 1–5				
	CV, %				
	Avg, >5				
	CV, %				
	Avg, All Conc		1,130	98,300	
	CV, %		34.2	46.3	
>4	Avg, <0.5		1,370[a]	124,000[a]	
	CV, %		21.5	35.3	
	Avg, 0.5 to <1				
	CV, %				
	Avg, 1–5				
	CV, %				
	Avg, >5				
	CV, %				
	Avg, All Conc		1,370	124,000	
	CV, %		21.5	35.3	

LS: loamy sand.
[a]Representative value.

ments are made, the quantum yield (5.46×10^{-3}) predicted field half-lives in the range of 3–10 d.

Puhl et al. (1983) employed a merry-go-round reactor with an artificial light source comprised of a medium-pressure Hg lamp to study photolysis. Irradiation was continuous but corrections were made to equate the irradiation to natural Kansas sunlight. The citation in Table 29 is based on equivalent Kansas summer sunlight.

The Gronberg (1984) study is the only one that used natural sunlight, and it indicates a difference in photolysis rate when data are plotted nonlinearly and linearly. This difference results from an artifact caused by sorption of cyfluthrin

Table 27. Cyfluthrin soil desorption partition coefficients.

Expt rating	Soil conc, ppm	Soil texture	K_{dd}	K_{ocd}	Reference
5	0.08	CL	1,740	69,900	Burhenne (1996)
5	0.1	LS	1,310	117,000	Burhenne (1996)
5	0.1	LS	974	141,000	Burhenne (1996)
5	0.1	SiL	1,450	161,000	Burhenne (1996)
All ratings	Avg, <0.5		1,360	122,000	
	CV, %		22.4	32.1	
	Avg, 0.5 to <1				
	CV, %				
	Avg, 1–5				
	CV, %				
	Avg, >5				
	CV, %				
	Avg, All Conc		1,360	122,000	
	CV, %		22.4	32.1	
>4	Avg, <0.5		1,360	122,000	
	CV, %		22.4	32.1	
	Avg, 0.5 to <1				
	CV, %				
	Avg, 1–5				
	CV, %				
	Avg, >5				
	CV, %				
	Avg, All Conc		1,360	122,000	
	CV, %		22.4	32.1	

Table 28. Hydrolysis of cyfluthrin in buffered water.

Tempera-ture, °C	pH	Nonlinear		Linear		Reference
		k^a	Half-life, d	k^a	Half-life, d	
25	5	0.00E + 00[b]	0	0.00E + 00	0	Sandie (1983)
25	7	3.79E – 03[b]	183	3.75E – 03	185	Sandie (1983)
25	9	3.77E – 01[b]	1.84	3.69E – 01	1.88	Sandie (1983)

[a] An entry of 0.00E + 00 means k was measured but cyfluthrin is stable.
[b] Representative value.

Table 29. Rate of photolysis of cyfluthrin in water.

Natural sun	Buffer medium	Nonlinear		Linear		Reference
		k	Half-life, d	k	Half-life, d	
No	Yes			2.31E – 01	3	Hellpointner (1990)
No	Yes			2.31E – 01	3	Hellpointner (1990)
No	Yes	5.86E – 02	11.8	5.68E – 02	12.2	Puhl et al. (1983)
Yes	Yes	*1.03E + 00*[a]	0.673	1.53E – 01	4.53	Gronberg (1984)

[a]Representative value.

to the walls of the exposure container, which caused the last 15% of parent to degrade much more slowly than the first 85%. For this reason, it is deemed that the rate constant from the nonlinear curve fitting is the best reflection of cyfluthrin photolysis rate, and it is this value that is chosen to be representative of cyfluthrin photolysis rate in water. All rate constants in Table 29 have been corrected for dark control reaction.

3. Photolysis on Soil.   Photolysis of cyfluthrin on soil surfaces was characterized by Puhl et al. (1983) in the laboratory under artificial light and again in 1986 by Chopade in an outdoor study under natural Kansas midsummer sunlight (Table 30). In both cases the moisture status of the soil during the experiment was air dry. Although there is little difference between nonlinear and linear rate data, the nonlinear rate constant under natural sunlight was selected as the representative value so that the standard practice of nonlinear selection throughout this review is maintained.

C. Biotic Chemical Properties

1. Aerobic Soil Degradation.   Wagner et al. (1983) examined cyfluthrin degradation in aerobic soils in the greenhouse at temperatures between 18° and 22 °C. (In this review, a mean value of 20 °C was assumed to represent incubation temperature.) Their application rate was 1 ppm, applied in toluene at 103 µL/ 50 g soil. Toluene is bacteriocidal and its presence in the experiment, with a

Table 30. Rate of photolysis of cyfluthrin on soil.

Natural sun	Moisture state	Nonlinear		Linear		Reference
		k	Half-life, d	k	Half-life, d	
No	Air dry	2.60E – 02	26.7	1.94E – 02	35.7	Puhl et al. (1983)
Yes	Air dry	*1.38E – 01*[a]	5.02	1.23E – 01	5.64	Chopade (1986)

[a]Representative value.

cyfluthrin rate of application bordering on being high, provided the reasons for a lower experimental rating of 3 (Table 31).

A second study was carried out by Yoshida et al. (1984) at a temperature of 28 °C and a lower cyfluthrin application of 0.5 ppm in 500 μL acetone/50 g soil. Although the study used nonradioactive material, extraction methodologies with aged samples were evaluated and analytical procedures appeared acceptable for providing quantitative data on parent material. Acetone volume is somewhat high, bordering on the realm of soil toxicity, but acetone is much less toxic than toluene to soil microorganisms and for this reason the experiment is rated higher than the study by Wagner et al. (1983).

2. Anaerobic Soil Degradation. Anaerobic degradation of cyfluthrin was studied by Wagner et al. (1983) utilizing the standard EPA anaerobic soil experimental design of 30 d aerobic followed by flooding with water plus N_2 atmosphere to cause anaerobicity (Table 32). Only two sampling periods were available for anaerobic kinetic evaluation, 30 and 60 d, and the first-order point equation (Eq. 4) was used to arrive at a cyfluthrin first-order rate constant under anaerobic soil conditions. Therefore, nonlinear estimations of rate are not available.

3. Aerobic Aquatic Degradation. No aerobic aquatic information was available for cyfluthrin.

4. Anaerobic Aquatic Degradation. No cyfluthrin anaerobic experiments were reviewed that utilized the anaerobic aquatic experimental design.

Table 31. Rate of cyfluthrin degradation in aerobic soil.

Expt rating	Temp, °C	Soil conc, ppm	Soil texture	Nonlinear		Linear		Reference
				k	Half-life, d	k	Half-life, d	
5	28	0.5	SiL	4.11E − 02	16.9	1.77E − 02	39.2	Yoshida et al. (1984)
5	28	0.5	L	1.12E − 01	6.2	2.41E − 02	28.8	Yoshida et al. (1984)
3	20	1	L	1.25E − 02	55.5	9.86E − 03	70.3	Wagner et al. (1983)
3	20	1	SL	1.12E − 02	61.7	7.88E − 03	88	Wagner et al. (1983)
	Avg, 0.5-<1.5				35.1		56.6	
	CV, %				78.8		48.4	
	Avg + CL				57.7		79.0	
>4	Avg				11.5[a]		34.0	
	CV, %				65.5		21.6	
	Avg + CL				28.0[a]		50.0	

[a]Representative value.

Table 32. Cyfluthrin anaerobic degradation generated by anaerobic soil experimental design.

Expt rating	Temp, °C	Soil conc, ppm	Soil texture	Nonlinear		Linear		Reference
				k	Half-life, d	k	Half-life, d	
7	18–22	1	SL			2.06E – 02	33.6	Wagner et al. (1983)
		Avg, <1.5 CV, %					33.6	
		Avg + CL					1.01E02	
>4		Avg CV, %					33.6[a]	
		Avg + CL					1.01E02[a]	

[a]Representative value.

V. Cypermethrin

Cypermethrin has a molecular weight of 416.3 and has been referred to as NRDC 149, PP383, WL 43467, LE 79–600, FMC 30980, OMS 2002, FMC 45806, and FMC Code 3765. Its Chemical Abstracts name is cyano(3-phenoxyphenyl)methyl 3 - (2,2 - dichloroethenyl) - 2,2 - dimethyl - cyclopropanecarboxylate [IUPAC = RS-α-cyano-3-phenoxybenzyl (1RS,3RS;1RS,3SR)-3-(2,2-dichlorovinyl)-2,2-dimethylcyclopropanecarboxylate]. Its CAS RN is 52315-07-8 and Official code is OMS 2002 (Tomlin 1994). Technical material has a cis/trans ratio of 52/48.

In this review the S isomer of cypermethrin, zetacypermethrin, is included under the cypermethrin heading; and the term cypermethrin represents zetacypermethrin as well as cypermethrin unless distinction is made between the two. Technical zetacypermethrin typically is 83% S and 89.7% total isomers. It has been referred to as FMC 56701, F56701, and F701. CAS RN is 52315-07-8, the same as cypermethrin (Tomlin 1994).

A. Physical Properties

1. Vapor Pressure. Grayson et al. (1982) employed a gas saturation technique to measure cypermethrin vapor pressure and measured the value shown in Table 33. Alvarez also used a gas saturation technique to measure the vapor pressures of cypermethrin (1991b) and zetacypermethrin (1991a) at 25 °C so that no extrapolation from vapor pressure at higher temperature was necessary. The Alvarez vapor pressure cited in Table 33 for cypermethrin (3.10×10^{-9} mm Hg) is the mean value from nine measurements at three flow rates, with a standard

Table 33. Cypermethrin vapor pressure.

Temperature, °C	Best value	VP, mm Hg	Source
20	Yes	1.40E − 09	Grayson et al. (1982)
25	Yes	3.10E − 09	Alvarez (1991b)
25	Yes	1.90E − 09	Alvarez (1991a)
25	Avg	2.5E − 09[a]	
	CV, %	33.9	

[a]Representative value.

deviation of 0.6×10^{-9} mm Hg; zetacypermethrin vapor pressure $(1.90 \times 10^{-9}$ mm Hg) is the average value from six measurements at two flow rates with standard deviation of the mean $= 0.6 \times 10^{-9}$ mm Hg. The data at 25 °C are selected to represent the vapor pressure of cypermethrin.

2. Water Solubility. Water solubility measurements summarized in Table 34 were performed by Wollerton and Husband (1988b) and again by Alvarez (1991b). Wollerton and Husband used a generator column coated with cypermethrin to generate saturated streams of buffer (pH 5, 7, and 9) that were analyzed for concentrations of cypermethrin. Values in Table 34 are the average values of six measurements made at each of the three pHs. Alvarez coated the walls of beakers with cypermethrin and then generated saturated solutions by continuous stirring with a magnetic stirrer for up to 5 d, with subsequent analysis of the water for cypermethrin content. The column saturation technique has less chance for suspension formation, and thus data from it are chosen to represent the water solubility of cypermethrin.

Table 34. Water solubility of cypermethrin.

Temperature, °C	Column saturation	WS, ppm	Reference
20	Yes	1.70E − 03	Wollerton and Husband (1988b)
20	Yes	4.00E − 03	Wollerton and Husband (1988b)
20	Yes	6.20E − 03	Wollerton and Husband (1988b)
20	Avg	3.97E − 03[a]	
	CV, %	56.7	
25	No	7.60E − 03	Alvarez (1991b)
25	No	4.50E − 02	Alvarez (1991b)
25	Avg	2.63E − 02	
	CV, %	100.6	

[a]Representative value.

3. Henry's Law Constant.　Assuming a molecular weight of 416.3, vapor pressure of 2.5×10^{-9} mm Hg, and water solubility of 4.0×10^{-3}, K_h from Eq. 1 = 3.4×10^{-7} atm m^3 mol^{-1}.

4. Octanol–Water Partition Coefficient.　Four references were reviewed with regard to the measurement of K_{ow} for cypermethrin (Table 35). The first study, by Wollerton and Husband (1988b), utilized a saturated column technique to generate a stream of water equilibrated with octanol and cypermethrin. The water stream was then analyzed for cypermethrin content.

　　Alvarez equilibrated octanol solutions of cypermethrin (1991b) and zetacypermethrin (1991a) with water in glass containers. Phases were sampled and analyzed for their cypermethrin and zetacypermethrin concentrations. Alvarez (1995) later revisited the measurement of K_{ow} with different methodology that minimized the formation of micelles and contamination of the water phase with octanol during sampling of the water. Table 35 summarizes this work, and the K_{ow} selected to represent cypermethrin is the mean of the data from the work of Wollerton and Husband (1988b) and Alvarez (1995). This value compares favorably with CLog P, the value for Log P calculated by Briggs (1999) from the molecular structure of cyfluthrin.

5. Bioconcentration Factor.　Baldwin and Lad (1978) exposed rainbow trout to cypermethrin in tanks treated each day with a single 1-d static renewal of chemical for up to 22 d. Mean concentration at the start of each day was 0.165 ppb and at the end 0.064 ppb, for an overall daily average of 0.115 ppb. Average fish concentration (total ^{14}C, whole-body basis) for days 11–22 was 0.0872 ppm, giving the BCF of 758 cited in Table 36.

　　Bennett (1981) also exposed rainbow trout to cypermethrin in a flow-through system that fed cypermethrin continuously into a fish aquarium. Water saturated by passage through a saturation generator column of nonradioactive cypermethrin was fed directly into the aquarium over an exposure period up to 18 d.

Table 35. Cypermethrin octanol–water partition coefficient.

Best values	K_{ow}	Log P	CLog P	Reference
Yes	3.98E + 06	6.60		Wollerton and Husband (1988b)
Yes	3.04E + 06	6.48		Alvarez (1995)
Avg	3.51E + 06	6.54[a]	6.1[b]	
CV, %	18.9	1.3		
No	6.46E + 02	2.81		Alvarez (1991b)
No	1.48E + 03	3.17		Alvarez (1991a)
Avg	1.06E + 03	2.99		
CV, %	55.5	8.5		

[a]Representative value.
[b]Calculated from molecular structure by Briggs (1999).

Table 36. Cypermethrin bioconcentration factor based on fish whole-body analysis.

Type of experiment	Best values	BCF	Reference
One-day renewal	No	758	Baldwin and Lad (1978)
Flow through	Yes	372	Giroir and Stuerman (1993)
Continuous flow	Yes	821	Bennett (1981)
	Avg	597[a]	
	CV, %	53.2	

[a]Representative value.

Water and fish were analyzed for cypermethrin, and BCF was calculated from cypermethrin mean water and fish concentrations. Giroir and Stuerman (1993) exposed bluegill sunfish continuously to ^{14}C-labeled cypermethrin up to 28 d. BCF was calculated from total ^{14}C concentrations in the water and fish at day 28 of exposure. In Table 36 the BCF of 597, selected to represent cypermethrin, is the average value of the measurements from the two flow-through studies.

6. Soil Sorption.   Two papers considered the sorption of cypermethrin to soil. The first was by Froelich (1991), who added cypermethrin in acetonitrile to dilute calcium chloride (2% acetonitrile cosolvent levels). Concentrations of 5–400 ppb (water solubility is 4 ppb) cypermethrin were prepared and equilibrated with soil in Teflon centrifuge tubes for 24 hr, samples were centrifuged, and water was analyzed for total ^{14}C content. Sorbed material was calculated by the difference method of ^{14}C in solution before and after equilibration with soil. Desorption was measured by replacement of solution with fresh reagent and reequilibration for another 24 hr, followed by the same analytical procedures used for sorption. The presence of cosolvent could have resulted in an underestimation of cypermethrin sorption properties and for this reason the Froelich study was given an experimental rating of 3.

Goggin et al. (1996) added cypermethrin in 100 μL acetone to 25 mL calcium chloride in glass centrifuge tubes containing different sediments sterilized with radiation before addition of cypermethrin. After 24 hr of equilibration, total ^{14}C in the soil and water phases were measured and used to calculate K_d, assuming all ^{14}C in both phases was cypermethrin. Analysis of the highest treatment rate indicated >97% of radioactivity in sediment and >60% in the water was cypermethrin. K_d was not corrected for these percentages, and therefore sorption is underestimated by an amount corresponding to the levels of radioactive noncypermethrin present in each sample. Desorption was measured by a similar procedure, after sorption reagent was removed and replaced with fresh calcium chloride. Adsorption data from both studies are summarized in Table 37, desorption data in Table 38, and Freundlich data in Table 39.

Table 37. Cypermethrin soil adsorption partition coefficients.

Expt rating	Soil conc, ppm	Soil texture	K_d	K_{oc}	Reference
3	0.07	SiL	70	2,690	Froelich (1991)
3	0.07	SiL	70	2,690	Froelich (1991)
3	0.09	S	90	39,100	Froelich (1991)
3	0.09	S	90	39,100	Froelich (1991)
3	0.1	SL	100	10,000	Froelich (1991)
3	0.1	CL	80	3,480	Froelich (1991)
3	0.1	CL	240	1,040	Froelich (1991)
6	0.5	SiL	16,500	526,000	Goggin et al. (1996)
6	0.5	L	23,900	180,000	Goggin et al. (1996)
6	0.5	SiL	2,200	223,000	Goggin et al. (1996)
3	0.6	SiL	149	5,720	Froelich (1991)
3	0.6	SiL	154	5,910	Froelich (1991)
3	0.7	CL	181	7,850	Froelich (1991)
3	0.7	CL	163	7,070	Froelich (1991)
3	0.7	SL	240	24,000	Froelich (1991)
3	0.7	SL	230	23,000	Froelich (1991)
3	0.8	S	217	94,200	Froelich (1991)
3	0.8	S	200	87,000	Froelich (1991)
6	1.5	L	19,800	149,000	Goggin et al. (1996)
6	1.6	SiL	16,200	518,000	Goggin et al. (1996)
3	2	SL	151	15,100	Froelich (1991)
3	2	SiL	238	9,140	Froelich (1991)
3	2	SL	257	25,700	Froelich (1991)
3	2	S	173	75,300	Froelich (1991)
3	2	S	239	104,000	Froelich (1991)
6	2	SiL	2,080	211,000	Goggin et al. (1996)
3	2	CL	238	10,400	Froelich (1991)
3	2	CL	229	9,940	Froelich (1991)
3	2	SiL	255	9,800	Froelich (1991)
6	3.9	SiL	15,600	498,000	Goggin et al. (1996)
3	4	S	298	129,000	Froelich (1991)
3	4	S	251	109,000	Froelich (1991)
3	4	SiL	250	9,620	Froelich (1991)
3	4	CL	222	9,640	Froelich (1991)
3	4	CL	224	9,730	Froelich (1991)
3	4	SL	285	28,500	Froelich (1991)
3	4	SiL	270	10,400	Froelich (1991)
3	4	SL	297	29,700	Froelich (1991)
6	4.5	SiL	2,200	223,000	Goggin et al. (1996)
6	4.5	L	30,800	232,000	Goggin et al. (1996)
3	8	S	371	161,000	Froelich (1991)
3	8	S	158	68,700	Froelich (1991)
3	8	SL	307	30,700	Froelich (1991)

Table 37. (Continued).

Expt rating	Soil conc, ppm	Soil texture	K_d	K_{oc}	Reference
3	8	CL	265	11,500	Froelich (1991)
3	8	CL	255	11,100	Froelich (1991)
3	8	SL	403	40,300	Froelich (1991)
3	9	SiL	350	13,500	Froelich (1991)
3	9	SiL	290	11,100	Froelich (1991)
6	12	SiL	2,910	295,000	Goggin et al. (1996)
6	13	L	19,600	147,000	Goggin et al. (1996)
6	13	SiL	14,600	466,000	Goggin et al. (1996)
All ratings	Avg, <0.5		106	14,000	
	CV, %		57.0	124.0	
	Avg, 0.5 to <1		4,010	108,000	
	CV, %		204.2	146.7	
	Avg, 1–5		4,120	110,000	
	CV, %		204.2	135.3	
	Avg, >5		3,590	114,000	
	CV, %		189.8	128.8	
	Avg, All Conc		3,430	97,300	
	CV, %		214.7	144.7	
>4	Avg, <0.5				
	CV, %				
	Avg, 0.5 to <1		14,200[a]	310,000[a]	
	CV, %		77.7	60.9	
	Avg, 1–5		14,400	305,000	
	CV, %		76.0	52.4	
	Avg, >5		12,400	303,000	
	CV, %		69.2	52.7	
	Avg, All Conc		13,900	306,000	
	CV, %		68.9	49.3	

[a]Representative value.

B. Abiotic Chemical Properties

1. Hydrolysis. Hydrolysis of cypermethrin was measured by Clifton (1992), who added cypermethrin to sterile buffer solutions in amber bottles capped with Teflon-lined lids. Bottles were incubated at 25 °C in the dark up to 30 d. Table 40 presents the first-order rate constants from nonlinear and linear analyses of datasets from buffers at pH 5, 7, and 9.

2. Photolysis in Water. Cypermethrin photolysis in water was studied by Estigoy et al. (1991b), who added radioactive material in acetonitrile to sterile buffer

Table 38. Cypermethrin soil desorption partition coefficients.

Expt rating	Soil conc, ppm	Soil texture	K_{dd}	K_{ocd}	Reference
3	0.06	SiL	100	3,850	Froelich (1991)
3	0.06	SiL	120	4,620	Froelich (1991)
3	0.08	S	160	69,600	Froelich (1991)
3	0.08	S	160	69,600	Froelich (1991)
3	0.09	SL	180	18,000	Froelich (1991)
3	0.09	SL	160	16,000	Froelich (1991)
3	0.1	CL	100	4,350	Froelich (1991)
3	0.1	CL	120	5,220	Froelich (1991)
6	0.5	SiL	15,000	480,000	Goggin et al. (1996)
6	0.5	SiL	2,940	298,000	Goggin et al. (1996)
6	0.5	L	30,700	231,000	Goggin et al. (1996)
3	0.5	SiL	156	5,980	Froelich (1991)
3	0.5	SiL	154	5,930	Froelich (1991)
3	0.6	SL	153	15,300	Froelich (1991)
3	0.6	CL	141	6,150	Froelich (1991)
3	0.6	CL	158	6,880	Froelich (1991)
3	0.6	SL	159	15,900	Froelich (1991)
3	0.7	S	106	46,300	Froelich (1991)
3	0.7	S	139	60,500	Froelich (1991)
3	2	CL	194	8,450	Froelich (1991)
3	2	SiL	206	7,930	Froelich (1991)
3	2	S	287	125,000	Froelich (1991)
3	2	SL	206	20,600	Froelich (1991)
3	2	CL	183	7,950	Froelich (1991)
6	2	SiL	2,380	242,000	Goggin et al. (1996)
6	2	SiL	17,800	569,000	Goggin et al. (1996)
3	2	SiL	279	10,700	Froelich (1991)
3	2	SL	179	17,900	Froelich (1991)
6	2	L	31,700	239,000	Goggin et al. (1996)
3	2	S	287	125,000	Froelich (1991)
3	3	CL	180	7,840	Froelich (1991)
3	3	CL	196	8,530	Froelich (1991)
3	4	S	534	232,000	Froelich (1991)
6	4	SiL	2,750	278,000	Goggin et al. (1996)
3	4	S	407	177,000	Froelich (1991)
3	4	SL	220	22,000	Froelich (1991)
3	4	SiL	247	9,480	Froelich (1991)
3	4	SiL	232	8,930	Froelich (1991)
3	4	SL	233	23,300	Froelich (1991)
6	5	L	10,700	80,300	Goggin et al. (1996)
6	5	SiL	20,000	639,000	Goggin et al. (1996)
3	7	SL	137	13,700	Froelich (1991)
3	7	CL	196	8,520	Froelich (1991)

Table 38. (Continued).

Expt rating	Soil conc, ppm	Soil texture	K_{dd}	K_{ocd}	Reference
3	7	CL	135	5,850	Froelich (1991)
3	7	SL	183	18,300	Froelich (1991)
3	8	SiL	242	9,310	Froelich (1991)
3	8	SiL	286	11,000	Froelich (1991)
3	8	S	486	211,000	Froelich (1991)
3	8	S	477	207,000	Froelich (1991)
6	13	L	23,500	177,000	Goggin et al. (1996)
6	13	SiL	3,020	306,000	Goggin et al. (1996)
6	13	SiL	20,000	638,000	Goggin et al. (1996)
All ratings	Avg, <0.5		138	23,900	
	CV, %		22.6	120.2	
	Avg, 0.5 to <1		4,530	107,000	
	CV, %		215.4	149.6	
	Avg, 1–5		2,940	107,000	
	CV, %		266.7	136.6	
	Avg, >5		6,100	179,000	
	CV, %		148.9	127.0	
	Avg, All Conc		3,630	112,000	
	CV, %		221.4	148.2	
>4	Avg, <0.5				
	CV, %				
	Avg, 0.5 to <1		16,200	336,000	
	CV, %		85.9	38.3	
	Avg, 1–5		13,700	332,000	
	CV, %		102.6	47.9	
	Avg, >5		15,400	368,000	
	CV, %		54.5	70.5	
	Avg, All Conc		15,000	348,000	
	CV, %		71.5	53.5	

and then irradiated samples of the solution in winter California sunlight up to 30 d. Dark controls were included and data from them were used to correct the data cited in Table 41 for dark reaction. Another study was conducted by Moffatt (1994a) in an apparatus designed specifically for quantum yield measurements. Light source was a xenon high-stability short arc lamp at 280 nm, and cypermethrin degradation rate was characterized periodically by high pressure liquid chromatography (HPLC). Dark controls were included to account for dark reaction. A quantum yield of 9.80×10^{-2} mmol meinstein^{-1} was calculated from the data, and the first-order rate constant cited in Table 41 was calculated from

Table 39. Freundlich sorption data for cypermethrin.

Expt rating	Soil texture	K_{df}	K_{ocf}	$1/n$	Reference
6	SiL	3,760	380,000	1.07	Goggin et al. (1996)
6	SiL	11,700	377,000	0.97	Goggin et al. (1996)
6	L	11,300	85,000	0.93	Goggin et al. (1996)
3	S	661	287,000	1.25	Froelich (1991)
3	SL	1,160	116,000	1.33	Froelich (1991)
3	SiL	1,900	72,400	1.47	Froelich (1991)
3	CL	416	18,300	1.15	Froelich (1991)
All ratings	Avg	4,410	191,000	1.17	
	CV, %	112.5	80.1	16.8	
>4	Avg	8,920	281,000	0.990	
	CV, %	50.1	60.4	7.3	

the quantum yield. The rate constant reflects the conditions of mid-European summer sun irradiating cypermethrin in a 5-cm layer of clean water. For purposes of property selection, the average of the two values generated under natural sunlight in Table 41 is chosen to represent the photolysis of cypermethrin in water.

3. Photolysis on Soil.   Estigoy et al. (1991a) studied the photolysis of cypermethrin on wet soil surfaces exposed to midwinter California sunlight up to 35 d. Kinetic results from nonlinear and linear first-order analyses of datasets from the study are cited in Table 42. Rate constants in the table are corrected for reaction in dark control samples included in the study.

Table 40. Hydrolysis of cypermethrin in buffered water.

Temperature, °C	pH	Nonlinear		Linear		Reference
		k[a]	Half-life, d[a]	k[a]	Half-life, d[a]	
25	5	*1.12E − 03*[b]	619	1.12E − 03	619	Clifton (1992)
25	7	*2.53E − 03*[b]	274	2.55E − 03	272	Clifton (1992)
25	9	*3.65E − 01*[b]	1.9	3.26E − 01	2.13	Clifton (1992)

[a]An entry of 0.00E + 00 means k was measured but cypermethrin is stable; a blank k or half-life value means no measurement.
[b]Representative value.

Table 41. Photolysis of cypermethrin in water.

Natural sun	Buffer medium	Nonlinear		Linear		
		k	Half-life, d	k	Half-life, d	Reference
No	Yes			1.11E + 00	0.625	Moffatt (1994a)
Yes	Yes	1.62E – 02	42.8	1.73E – 02	40.1	Estigoy et al. (1991b)
Yes	Yes	2.98E – 02	23.3	3.39E – 02	20.4	Estigoy et al. (1991b)
Avg		*2.30E – 02*[a]				
CV, %		41.8				

[a]Representative value.

C. Biotic Chemical Properties

1. Aerobic Soil Degradation. Studies on the degradation of cypermethrin in aerobic soils (Table 43) have been performed by Swaine and Hayward (1979), Harvey et al. (1981), and by Ramsey (1991a). Swaine and Hayward added cypermethrin in 500 μL acetone to 100 g soil in conical flasks that were then incubated according to German pesticide registration guidelines. Harvey et al. applied cypermethrin in 100 μL acetone to the surface of several soils and incubated them without further mixing in a continuously aerated glass system. Ramsey applied cypermethrin in fairly high concentrations (3 ppm) in quantities of ethanol equal to 500 μL/50 g of soil. Incubations were in the dark up to 150 d. Results from the nonlinear and linear analyses of datasets from all three studies are summarized in Table 43, along with the ratings assigned to each experimental design. Data from Ramsey et al. were given a rating of lesser confidence on the basis of higher application rate coupled with the use of a large volume of ethanol as solvent.

2. Anaerobic Soil Degradation. Ramsey (1991b) utilized an anaerobic soil experimental design to study the anaerobic degradation of cypermethrin. Cypermethrin in 500 μL ethanol was added dropwise with mixing to 50 g soil in soil biometer flasks and then incubated aerobically in the dark for 32 d before flooding with 60 mL water and amendment with cellulose to ensure anaerobicity.

Table 42. Photolysis of cypermethrin on soil.

Natural sun	Moisture state	Nonlinear		Linear		
		k	Half-life, d	k	Half-life, d	Reference
Yes	75% 1/3 bar	*4.19E – 03*[a]	165	4.32E – 03	160	Estigoy et al. (1991a)

[a]Representative value.

Table 43. Cypermethrin degradation in aerobic soil.

Expt rating	Temp, °C	Soil conc, ppm	Soil texture	Nonlinear		Linear		Reference
				k	Half-life, d	k	Half-life, d	
8	15	0.6	CL	4.28E − 02	16.2	1.78E − 02	38.9	Harvey et al. (1981)
8	25	0.6	Fen peat	2.62E − 02	26.5	1.46E − 02	47.5	Harvey et al. (1981)
8	25	0.6	LCS	2.45E − 02	28.3	1.43E − 02	48.5	Harvey et al. (1981)
8	25	0.6	CL	1.08E − 01	6.42	4.27E − 02	16.2	Harvey et al. (1981)
5	22	1		3.49E − 02	19.9	1.73E − 02	40.1	Swaine and Hayward (1979)
5	22	1		9.50E − 03	73	6.47E − 03	107	Swaine and Hayward (1979)
3	25	3	FSL	1.32E − 02	52.5	1.15E − 02	60.3	Ramsey (1991a)
8	25	6	CL	3.01E − 02	23	1.91E − 02	36.3	Harvey et al. (1981)
All ratings		Avg, <0.5						
		CV, %						
		Avg + CL						
		Avg, 0.5–1.5			28.4		49.7	
		CV, %			81.8		61.2	
		Avg + CL			42.4		68.0	
		Avg, 1.5–10			37.8		48.3	
		CV, %			55.3		35.1	
		Avg + CL			83.2		85.2	
		Avg, >10						
		CV, %						
		Avg + CL						
		Avg, All Conc			30.7		49.4	
		CV, %			70.3		53.7	
		Avg + CL			41.5		49.4	
>4		Avg, All Conc			27.6[a]		47.8	
		CV, %			77.1		59.0	
		Avg + CL			39.2[a]		63.1	

LCS: loamy coarse sand; FSL: fine sandy loam.
[a]Representative value.

Anaerobic incubation in the dark continued up to an additional 60 d. Nonlinear and linear kinetic analyses of data from the study are summarized in Table 44.

3. Aerobic Aquatic Degradation.   Degradation of cypermethrin in aerobic aquatic systems was studied by Rapley et al. (1981), Elmarakby (1998), and Lucas

(1998). Rapley added cypermethrin in 100 μL acetone to natural water and sediment samples in cylinders aerated continuously and incubated in the dark at 16 °C up to 182 d. Nonlinear and linear analyses of results from the Rapley study are presented in Table 45. Analyses utilized only the data through 35 d because most material added to each system (90% or more) had been degraded by then.

Elmarakby (1998) added material as zetacypermethrin in 50–100 μL acetonitrile to 100 mL water plus sediment collected from a rice-growing area in California. Incubation containers were Nalgene bottles aerated continuously and incubated in the dark for up to 30 d at 25 °C. The third study, by Lucas (1998), added zetacypermethrin in 70–120 μL acetonitrile to natural water–sediment systems equilibrated in glass cylinders for 71 d before treatment. The cylinders were aerated continuously in the dark at 20 °C and were sampled periodically up to an additional 99 d. Nonlinear and linear analyses of datasets from each of the three studies are summarized in Table 45.

4. Anaerobic Aquatic Degradation. Ramsey (1998) utilized an anaerobic aquatic experimental design to evaluate the degradation of cypermethrin in anaerobic systems. He applied zetacypermethrin in 100 μL acetonitrile to the water

Table 44. Cypermethrin degradation generated by anaerobic soil experimental design.

Expt rating	Temp, °C	Soil conc, ppm	Soil texture	Nonlinear		Linear		Reference
				k	Half-life, d	k	Half-life, d	
5	25	3	FSL	1.26E − 02	55	1.23E − 02	56.4	Ramsey (1991b)
All ratings		Avg, <1.5 CV, % Avg + CL						
		Avg, >1.5 CV, % Avg + CL			55 165		56.4 169	
		Avg, All Conc CV, % Avg + CL			55 165		56.4 169	
>4		Avg, All Conc CV, % Avg + CL			*55*[a] *165*[a]		56.4 169	

[a]Representative value.

D.A. Laskowski

Table 45. Cypermethrin degradation in aerobic aquatic systems.

Sample name	Sediment conc, ppm	Nonlinear		Linear		Reference
		k	Half-life, d	k	Half-life, d	
Sediment	0.75	1.12E − 01	6.19	7.36E − 02	9.42	Elmarakby (1998)
Mill Stream	0.08	1.86E − 01	3.73	3.82E − 02	18.1	Lucas (1998)
Iron Hatch	0.08	2.37E − 01	2.92	3.97E − 02	17.5	Lucas (1998)
Kennet	0.2	8.55E − 02	8.11	6.41E − 02	10.8	Rapley et al. (1981)
Loddon	0.2	1.05E − 01	6.6	6.30E − 02	11	Rapley et al. (1981)
Whitewater	0.2	1.56E − 01	4.44	6.59E − 02	10.5	Rapley et al. (1981)
Warfield	0.2	3.44E − 02	20.1	3.19E − 02	21.7	Rapley et al. (1981)
Avg			7.44[a]		14.1	
CV, %			78.8		34.2	
Avg + CL			10.6[a]		16.8	

[a]Representative value.

Table 46. Cypermethrin anaerobic degradation generated by anaerobic aquatic experimental design.

Expt rating	Temp, °C	Soil conc, ppm	Soil texture	Nonlinear		Linear		Reference
				k	Half-life, d	k	Half-life, d	
8	25	0.3	CL	9.37E − 02	7.4	2.23E − 02	31.1	Ramsey (1998)
All ratings		Avg, <1.5 CV, %			7.4		31.1	
		Avg + CL			22.2		93.3	
		Avg, >1.5 CV, %						
		Avg + CL						
		Avg, All Conc CV, %			7.4		31.1	
		Avg + CL			22.2		93.3	
>4		Avg, All Conc CV, %			7.4[a]		31.1	
		Avg + CL			22.2[a]		93.3	

[a]Representative value.

layer of a biometer flask containing 80 mL water and 40 g sediment collected from a rice-growing area of California. Samples had been preincubated in the dark at 25 °C under nitrogen gas for 30 d before treatment. After treatment, samples were incubated again at 25 °C in the dark and under a nitrogen atmosphere for up to 183 d. Results of nonlinear and linear analyses of a composite dataset arising from experiments with two ^{14}C labels are presented in Table 46.

VI. Deltamethrin and Tralomethrin

Deltamethrin [C.A. = [1 R-[1α(S*),3α]] - cyano(3 - phenoxyphenyl)methyl 3-(2,2-dibromoethenyl) - 2,2 - dimethylcyclopropanecarboxylate (IUPAC = (S) - α-cyano-3-phenoxybenzyl (1R,3R)-3-(2,2-dibromovinyl)-2,2-dimethylcyclopropanecarboxylate)] has a molecular weight of 505.2. It has been referred to as decamethrin, RU 22974, FMC 45498, Decis®, K-Othrin®, Cislin®, Butox®, and Rup 987. CAS RN is 52820-00-5; it also has a development code of NRDC 161 and official code of OMS 1998 (Tomlin 1994).

  Tralomethrin is included with deltamethrin because it is a deltamethrin analog and degrades to deltamethrin. Its C.A. name is cyano(3-phenoxyphenyl)-

Table 47. Vapor pressure of deltamethrin.

Temperature, °C	Best value	VP, mm Hg	Source
25	No	1.50E – 08	Grelet (1990)
25	No	3.00E – 10	Agrevo b (unknown date)
Avg		7.6E – 09	
CV, %		135.9	
25	Yes	*9.32E – 11*[a]	Yoder (1991a)
35	No	3.10E – 10	Yoder (1991a)
45	No	1.49E – 09	Yoder (1991a)

[a]Representative value.

methyl 2,2-dimethyl-3-(1,2,2,2-tetrabromoethyl)cyclopropanecarboxylate [IU-PAC = (S)-α-cyano-3-phenoxybenzyl (1R,3S)-2,2-dimethyl-3-[(RS)-1,2,2,2-tet-rabromoethyl]cyclopropanecarboxylate]. Its molecular weight is 665.0, and it has been called RU 25474, NU 831, HAG 107, and Scout®. CAS RN is 66841–25–6; the official code is OMS 3048.

A. Physical Properties

1. Vapor Pressure. Vapor pressure of deltamethrin has been measured on three different occasions. In Table 47 the first value cited was from Grelet (1990), but this reference is a summary and describes no methodology or cites any actual data. For this reason, the value is considered questionable and is not chosen as best value.

The second reference, from Agrevo, year unknown, is a partial report on the physical properties of tralomethrin. However, it contained an attachment describing the measurement of deltamethrin vapor pressure at elevated temperatures (156°–187 °C) via a gas saturation method. Vapor pressure at 25 °C was obtained by graphical extrapolation from data at the higher temperatures, and therefore this value also was not chosen as a best value for deltamethrin. Measurements were made a third time in 1991 (Yoder 1991a), using the gas saturation technique and temperatures of 25°, 35°, and 45 °C. The value at 25 °C is a

Table 48. Vapor pressure of tralomethrin.

Temperature, °C	Best value	VP, mm Hg	Source
25	No	1.30E – 13	Agrevo b (unknown date)
25	No	3.60E – 11	Tillier (1993)
Avg		*1.8E – 11*[a]	
CV, %		140.4	
20	No	1.50E – 11	Tillier (1993)

[a]Representative value.

Table 49. Water solubility of deltamethrin.

Temperature, °C	Column saturation	WS, ppm	Reference
20	No	2.00E − 04	Grelet (1990)
Avg		2.00E − 04[a]	
CV, %			

[a]Representative value.

direct measurement, not an extrapolation, and thus is considered to be the best representation of deltamethrin vapor pressure.

Tralomethrin vapor pressure (Table 48) was measured in the Agrevo reference just described for deltamethrin (Agrevo b, unknown date), but the measurement is not deemed a "best value" because of the same temperature extrapolation issue. Measurements were carried out again by Tillier (1993), using a gas saturation method and temperatures of 60°, 70°, and 98 °C, with subsequent extrapolation to 25° and 20 °C from the higher temperature data. Because extrapolation was used, data again were not deemed "best value." However, no other data were available, and it was decided to average the two values at 25 °C in Table 48 and use this average to represent tralomethrin vapor pressure.

2. Water Solubility.   Deltamethrin solubility in water at 20 °C has been reported by Grelet (1990). This reference is a summary and provides no description of methodology or citation of experimental data. The value cited in Table 49 was taken directly from this reference.

Table 50 summarizes the water solubility information for tralomethrin. The first solubility was obtained from the partial company report (Agrevo b, unknown date) discussed earlier. The second value was developed from measurements by Pepin and Gargot (1988), who used a stirred vessel methodology to obtain saturated solutions. The value cited in the table is the mean of three measurements, but it could be an overestimate of solubility because the stirred vessel technique is susceptible to formation of suspensions that create an excess of chemical in the water phase. No other data are available, however, so this value is selected to represent the water solubility of tralomethrin.

Table 50. Water solubility of tralomethrin.

Temperature, °C	Column saturation	WS, ppm	Reference
20	No	7.10E − 02	Agrevo b (unknown date)
25	No	8.40E − 02[a]	Pepin and Gargot (1988)

[a]Representative value.

Table 51. Deltamethrin octanol–water partition coefficient.

Best values	K_{ow}	Log P	CLog P	Reference
No	2.73E + 05	5.44		Grelet (1990)
Yes	3.42E + 04	4.53[a]	6.5[b]	Yoder (1991b)

[a]Representative value.
[b]Calculated from molecular structure by Briggs (1999).

3. Henry's Law Constant.   Calculation of K_h from Eq. 1 assuming vapor pressures of 9.3×10^{-11} mm Hg for deltamethrin, 1.8×10^{-11} for tralomethrin, water solubilities of 2.0×10^{-4} ppm for deltamethrin and 8.4×10^{-2} ppm for tralomethrin, and a molecular weight of 505.2 for deltamethrin and 665 for tralomethrin provides these values: deltamethrin K_h, 3.1×10^{-7} atm m^3 mol^{-1}, and tralomethrin K_h, 1.9×10^{-10} atm m^3 mol^{-1}.

4. Octanol–Water Partition Coefficient. K_{ow} measurements for deltamethrin were made by Grelet (1990) and by Yoder (1991b) using phase equilibrium methodology (Table 51). Water and octanol phases containing deltamethrin were equilibrated with each other and then sampled to measure deltamethrin concentrations in both phases. Although the procedure is susceptible to underestimation of K_{ow} because of the potential for emulsion formation and contamination of water phase by octanol during sampling of the water, measurements by Yoder included careful attention to the prevention and detection of emulsions. The value from Yoder (1991b) is selected to represent K_{ow} for deltamethrin. Comparison to CLog P calculated from deltamethrin molecular structure by Briggs (1999) indicates the calculated value is substantially higher.

Table 52 gives K_{ow} values for tralomethrin. The first value in the table was extracted from the summary report (Agrevo b, unknown date) of tralomethrin properties discussed earlier, and the second came from the work of Gargot (1994). Gargot used the phase equilibration methodology with stirring to measure distributions of tralomethrin between octanol and water. The Gargot value reported in Table 52 is the mean of six measurements and is selected to represent K_{ow} for tralomethrin because of the summary nature of the report referenced as Agrevo b, (unknown date). The Gargot value possibly underestimates K_{ow}

Table 52. Tralomethrin octanol–water partition coefficient.

Best values	K_{ow}	Log P	CLog P	Reference
No	1.00E + 05	5.00		Agrevo b (unknown date)
Yes	1.19E + 05	5.05[a]	7.6[b]	Gargot (1994)

[a]Representative value.
[b]Calculated from molecular structure by Briggs (1999).

because of the issues related to the use of phase equilibrium methodology for very nonpolar chemicals such as pyrethroids. Comparison of it to CLog P in the table indicates that CLog P is higher.

5. Bioconcentration Factor.   Bioconcentration of deltamethrin (Table 53) by fish exposed continuously in flow-through aquaria was measured by Fackler in 1990 and again in 1993 by Schocken. The Fackler study did not identify ^{14}C material found in fish because too little tissue was available to provide enough ^{14}C-labeled material for analysis, and so BCF calculation was based on total ^{14}C in the water and fish. The Schocken study performed fish tissue analysis, showing that 80%–85% of the ^{14}C in the fish was deltamethrin. BCF was not corrected for this, and thus the Schocken value in Table 53 overestimates deltamethrin BCF. The Schocken value is selected to represent deltamethrin BCF, however, because of the additional identification work that the Fackler experiment lacked.

Bioconcentration information for tralomethrin was not available and is not included in this review.

6. Soil Sorption.   Two studies of deltamethrin sorption to soil (Table 54) have been carried out. The first by Smith (1990b) used ^{14}C-labeled deltamethrin added in <0.2% acetonitrile to dilute calcium chloride solutions (four different deltamethrin concentrations) contained in glass tubes with soil and (tubes) pretreated with nonradioactive deltamethrin to minimize wall adsorption effects. Total radioactivity remaining in solution after equilibration was measured, and chemical sorbed to soil was calculated by the difference method after correction for wall adsorption with soil-less controls. K_d and K_{oc} were then calculated from the water-phase and soil-phase deltamethrin concentrations.

Desorption was measured by decantation of calcium chloride from sorption samples, addition of fresh calcium chloride, reequilibration, and then measurement of water phase for total ^{14}C followed by calculation of soil concentration by the difference method.

The second study, by Christensen (1993), used similar methodology with no more than 0.1% acetone present in the calcium chloride. Desorption measurements were performed on sorption samples by decantation and replenishment of calcium chloride with fresh reagent. Results for adsorption are presented in Table 54, desorption in Table 55, and Freundlich sorption data in Table 56.

Table 53. Deltamethrin bioconcentration factor based on fish whole-body analysis.

Type of experiment	Best values	BCF	Reference
Flow-through	Yes	698[a]	Schocken (1993)
Flow-through	No	1400	Fackler (1990)

[a]Representative value.

Table 54. Deltamethrin soil adsorption partition coefficients.

Expt rating	Soil conc, ppm	Soil texture	K_d	K_{oc}	Reference
7	0.02	SiL	377	50,300	Christensen (1993)
7	0.02	SiL	511	68,100	Christensen (1993)
7	0.04	SL	1,540	262,000	Christensen (1993)
7	0.04	SiL	1,820	243,000	Christensen (1993)
7	0.04	SL	2,510	545,000	Christensen (1993)
7	0.04	SL	1,670	362,000	Christensen (1993)
7	0.04	SiL	2,250	278,000	Christensen (1993)
7	0.05	SiL	3,460	427,000	Christensen (1993)
7	0.05	SiL	2,670	330,000	Christensen (1993)
7	0.08	SL	1,200	262,000	Christensen (1993)
7	0.09	SL	1,360	296,000	Christensen (1993)
7	0.09	SL	1,410	306,000	Christensen (1993)
7	0.1	SiL	3,020	403,000	Christensen (1993)
7	0.1	SiL	2,490	308,000	Christensen (1993)
7	0.1	SiL	3,680	454,000	Christensen (1993)
7	0.1	SiL	2,250	300,000	Christensen (1993)
7	0.1	SiL	2,110	281,000	Christensen (1993)
7	0.1	SiL	3,600	445,000	Christensen (1993)
7	0.2	SL	1,300	283,000	Christensen (1993)
7	0.2	SiL	2,380	294,000	Christensen (1993)
7	0.2	SiL	4,100	506,000	Christensen (1993)
7	0.2	SL	1,160	253,000	Christensen (1993)
7	0.2	SL	1,170	255,000	Christensen (1993)
7	0.2	SiL	3,160	421,000	Christensen (1993)
7	0.2	SiL	3,910	521,000	Christensen (1993)
7	0.3	SL	1,150	249,000	Christensen (1993)
7	0.3	SL	1,130	246,000	Christensen (1993)
7	0.3	SiL	4,930	658,000	Christensen (1993)
7	0.4	SL	1,600	347,000	Christensen (1993)
7	0.4	SiL	1,980	264,000	Christensen (1993)
7	0.5	SiL	3,130	387,000	Christensen (1993)
7	0.5	SiL	4,690	626,000	Christensen (1993)
7	0.5	SL	1,280	2,210,000	Smith (1990b)
7	0.5	SL	860	1,480,000	Smith (1990b)
7	0.5	SL	284	490,000	Smith (1990b)
7	0.5	SL	1,100	478,000	Smith (1990b)
7	0.5	SL	1,100	478,000	Smith (1990b)
7	0.5	SL	1,130	491,000	Smith (1990b)
7	0.5	SL	1,010	1,740,000	Smith (1990b)
7	0.5	C	1,230	535,000	Smith (1990b)
7	0.5	C	1,230	535,000	Smith (1990b)
7	0.5	SiL	3,770	466,000	Christensen (1993)
7	0.5	SiL	4,750	587,000	Christensen (1993)
7	0.5	SiL	2,590	345,000	Christensen (1993)

Table 54. (Continued).

Expt rating	Soil conc, ppm	Soil texture	K_d	K_{oc}	Reference
7	0.6	C	1,470	639,000	Smith (1990b)
7	0.7	SL	494	215,000	Smith (1990b)
7	0.7	SiCL	4,030	498,000	Smith (1990b)
7	0.8	SiCL	5,170	638,000	Smith (1990b)
7	0.8	SiCL	4,300	531,000	Smith (1990b)
7	1	SL	1,220	2,100,000	Smith (1990b)
7	1	SL	1,550	2,670,000	Smith (1990b)
7	1	SL	2,090	909,000	Smith (1990b)
7	1	C	1,930	839,000	Smith (1990b)
7	1	C	2,550	1,110,000	Smith (1990b)
7	1	C	1,640	713,000	Smith (1990b)
7	1	SiCL	4,470	552,000	Smith (1990b)
7	1	SL	2,060	896,000	Smith (1990b)
7	2	SiCL	3,300	407,000	Smith (1990b)
7	2	SiCL	4,360	538,000	Smith (1990b)
7	3	SL	2,040	887,000	Smith (1990b)
7	3	SL	1,410	2,430,00	Smith (1990b)
7	3	SiCL	3,460	427,000	Smith (1990b)
7	3	SiCL	3,970	490,000	Smith (1990b)
7	3	SiCL	3,340	412,000	Smith (1990b)
7	3	C	2,810	1,220,000	Smith (1990b)
7	3	C	4,280	1,860,000	Smith (1990b)
7	3	C	2,500	1,090,000	Smith (1990b)
7	3	SL	2,680	1,170,000	Smith (1990b)
7	3	SL	1,810	3,120,000	Smith (1990b)
7	3	SL	2,260	983,000	Smith (1990b)
7	3	SL	2,140	3,690,000	Smith (1990b)
7	3	SL	2,760	1,200,000	Smith (1990b)
7	3	SL	2,480	1,080,000	Smith (1990b)
7	4	SL	1,640	713,000	Smith (1990b)
7	5	SL	1,200	2,070,000	Smith (1990b)
7	5	SL	1,470	639,000	Smith (1990b)
7	5	SL	1,990	865,000	Smith (1990b)
7	5	SL	1,950	848,000	Smith (1990b)
7	6	SL	1,940	3,340,000	Smith (1990b)
7	6	SL	1,880	3,240,000	Smith (1990b)
7	6	SL	1,340	2,310,000	Smith (1990b)
7	6	SL	1,910	3,290,000	Smith (1990b)
7	6	C	1,820	791,000	Smith (1990b)
7	6	C	1,800	783,000	Smith (1990b)
7	7	C	2,940	1,280,000	Smith (1990b)
7	7	SiCL	5,890	727,000	Smith (1990b)
7	7	SiCL	5,740	709,000	Smith (1990b)
7	7	C	2,550	1,110,000	Smith (1990b)

Table 54. (Continued).

Expt rating	Soil conc, ppm	Soil texture	K_d	K_{oc}	Reference
7	7	SL	2,430	4,190,000	Smith (1990b)
7	7	SiCL	5,870	725,000	Smith (1990b)
7	7	SiCL	4,470	552,000	Smith (1990b)
7	8	C	3,300	1,430,000	Smith (1990b)
7	8	C	3,320	1,440,000	Smith (1990b)
7	8	SiCL	3,960	489,000	Smith (1990b)
7	8	SiCL	5,550	685,000	Smith (1990b)
All	Avg, <0.5		2,200	331,000	
ratings	CV, %		51.2	38.9	
	Avg, 0.5 to <1		2,300	704,000	
	CV, %		72.0	73.5	
	Avg, 1–5		2,590	1,260,000	
	CV, %		36.7	70.4	
	Avg, >5		3,020	1,500,000	
	CV, %		53.1	74.6	
	Avg, All Conc		2,500	908,000	
	CV, %		53.5	95.3	
>4	Avg, <0.5		2,200	331,000	
	CV, %		51.2	38.9	
	Avg, 0.5 to <1		2,300[a]	704,000[a]	
	CV, %		72.0	73.5	
	Avg, 1–5		2,590	1,260,000	
	CV, %		36.7	70.4	
	Avg, >5		3,020	1,500,000	
	CV, %		53.1	74.6	
	Avg, All Conc		2,500	908,000	
	CV, %		53.5	95.3	

[a]Representative value.

Tralomethrin sorption and desorption has been measured by Daly (1989) by equilibration of soils with ^{14}C-tralomethrin solutions containing up to 3.4% acetonitrile (% not held constant) in dilute calcium chloride. After equilibration, total ^{14}C remaining in the water phase was measured, and soil-phase total ^{14}C concentration was then calculated by the difference method. TLC assay of the water phase after equilibration indicated the presence of ^{14}C material other than tralomethrin (15%–52%), but partition coefficients were not corrected for this. Desorption was measured by decantation and readdition of fresh calcium chloride reagent to sorption samples after equilibration. Water- and soil-phase concentrations were determined by the same procedures used for sorption.

Table 55. Deltamethrin soil desorption partition coefficients.

Expt rating	Soil conc, ppm	Soil texture	K_{dd}	K_{ocd}	Reference
7	0.1	SiL	1,550	191,000	Christensen (1993)
7	0.2	SiL	2,560	341,000	Christensen (1993)
7	0.2	SiL	3,180	424,000	Christensen (1993)
7	0.2	SiL	3,770	503,000	Christensen (1993)
7	0.2	SiL	2,600	321,000	Christensen (1993)
7	0.2	SiL	1,550	191,000	Christensen (1993)
7	0.2	SiL	3,190	394,000	Christensen (1993)
7	0.2	SL	1,520	330,000	Christensen (1993)
7	0.2	SL	1,740	378,000	Christensen (1993)
7	0.3	SL	4,200	913,000	Christensen (1993)
7	0.3	SiL	2,250	300,000	Christensen (1993)
7	0.3	SiL	2,560	341,000	Christensen (1993)
7	0.3	SiL	4,970	614,000	Christensen (1993)
7	0.3	SiL	6,180	763,000	Christensen (1993)
7	0.3	SiL	2,090	279,000	Christensen (1993)
7	0.3	SL	1,600	130,000	Christensen (1993)
7	0.3	SL	1,960	426,000	Christensen (1993)
7	0.4	SL	3,280	713,000	Christensen (1993)
7	2	SL	1,830	3,160,000	Smith (1990b)
7	3	SL	2,760	1,200,000	Smith (1990b)
7	3	SL	3,470	1,510,000	Smith (1990b)
7	3	SL	3,200	1,390,000	Smith (1990b)
7	4	SL	3,150	5,430,000	Smith (1990b)
7	4	SL	3,450	5,950,000	Smith (1990b)
7	4	C	4,090	1,780,000	Smith (1990b)
7	5	C	4,590	2,000,000	Smith (1990b)
7	5	C	4,100	1,780,000	Smith (1990b)
7	5	SiCL	8,360	1,030,000	Smith (1990b)
7	5	SiCL	6,460	798,000	Smith (1990b)
7	6	SiCL	9,130	1,130,000	Smith (1990b)
All ratings	Avg, <0.5		2,820	420,000	
	CV, %		46.0	50.0	
	Avg, 0.5 to <1				
	CV, %				
	Avg, 1–5		3,140	2,920,000	
	CV, %		22.4	68.7	
	Avg, >5		6,530	1,350,000	
	CV, %		34.1	38.3	
	Avg, All Conc		3,510	1,160,000	
	CV, %		54.6	121.8	

Table 55. (Continued).

Expt rating	Soil conc, ppm	Soil texture	K_{dd}	K_{ocd}	Reference
>4	Avg, <0.5		2,820	420,000	
	CV, %		46.0	50.0	
	Avg, 0.5 to <1				
	CV, %				
	Avg, 1–5		3,140	2,920,000	
	CV, %		22.4	68.7	
	Avg, >5		6,530	1,350,000	
	CV, %		34.1	38.3	
	Avg, All Conc		3,510	1,160,000	
	CV, %		54.6	121.8	

Tralomethrin sorption data are presented in Table 57, desorption in Table 58, and Freundlich sorption data in Table 59.

B. Abiotic Chemical Properties

1. Hydrolysis. Deltamethrin hydrolysis at pH 5, 7, and 9 was measured by Smith (1990a) using ^{14}C-labeled chemical added to sterile buffers incubated in the dark at 25 °C. Nonlinear and linear first-order rate data are presented in Table 60 for each pH value.

Two studies with tralomethrin have been performed. One study (Agrevo a,

Table 56. Freundlich sorption data for deltamethrin.

Expt rating	Soil texture	K_{df}	K_{ocf}	$1/n$	Reference
7	SiL	5,260	701,000	1.33	Christensen (1993)
7	SiL	4,280	528,000	1.09	Christensen (1993)
7	SL	929	202,000	0.83	Christensen (1993)
7	SL	3,950	6,810,000	1.17	Smith (1990b)
7	SL	8,890	3,870,000	1.25	Smith (1990b)
7	C	16,900	7,370,000	1.29	Smith (1990b)
7	SiCL	3,440	425,000	0.975	Smith (1990b)
All ratings	Avg	6,240	2,840,000	1.13	
	CV, %	84.5	113.0	16.0	
>4	Avg	6,240	2,840,000	1.13	
	CV, %	84.5	113.0	16.0	

Table 57. Tralomethrin soil adsorption partition coefficients.

Expt rating	Soil conc, ppm	Soil texture	K_d	K_{oc}	Reference
3	0.2	SL	1,150	288,000	Daly (1989)
3	0.2	S	680	272,000	Daly (1989)
3	0.2	S	1,100	440,000	Daly (1989)
3	0.2	SL	984	246,000	Daly (1989)
3	0.2	CL	1,370	105,000	Daly (1989)
3	0.2	L	1,470	327,000	Daly (1989)
3	0.2	L	1,460	324,000	Daly (1989)
3	0.2	CL	1,280	985,000	Daly (1989)
3	0.9	SL	1,350	338,000	Daly (1989)
3	0.9	S	677	271,000	Daly (1989)
3	0.9	S	925	370,000	Daly (1989)
3	0.9	CL	1,550	119,000	Daly (1989)
3	0.9	SL	1,400	350,000	Daly (1989)
3	0.9	CL	1,650	127,000	Daly (1989)
3	0.9	L	1,650	367,000	Daly (1989)
3	0.9	L	1,880	418,000	Daly (1989)
3	1.5	CL	1,830	141,000	Daly (1989)
3	1.5	L	1,310	291,000	Daly (1989)
3	1.5	S	1,150	460,000	Daly (1989)
3	1.5	L	1,410	313,000	Daly (1989)
3	1.5	S	656	262,000	Daly (1989)
3	1.5	SL	1,480	370,000	Daly (1989)
3	1.5	CL	1,810	138,000	Daly (1989)
3	1.5	SL	1,410	353,000	Daly (1989)
3	2	S	1,200	480,000	Daly (1989)
3	2	CL	1,930	148,000	Daly (1989)
3	2	SL	1,450	363,000	Daly (1989)
3	2	L	1,690	376,000	Daly (1989)
3	2	L	473	105,000	Daly (1989)
3	2	CL	2,020	155,000	Daly (1989)
3	2	S	995	398,000	Daly (1989)
3	2	SL	1,230	308,000	Daly (1989)
All ratings	Avg, <0.5		1,190	263,000	
	CV, %		22.6	43.7	
	Avg, 0.5 to <1		1,390[a]	295,000[a]	
	CV, %		29.0	38.6	
	Avg, 1–5		1,380	291,000	
	CV, %		31.5	41.5	
	Avg, >5				
	CV, %				
	Avg, All Conc		1,330	285,000	
	CV, %		29.2	40.2	

Table 57. (Continued).

Expt rating	Soil conc, ppm	Soil texture	K_d	K_{oc}	Reference
>4	Avg, <0.5				
	CV, %				
	Avg, 0.5 to <1				
	CV, %				
	Avg, 1–5				
	CV, %				
	Avg, >5				
	CV, %				
	Avg, All Conc				
	CV, %				

[a]Representative value.

unknown date) was conducted at concentrations (10 and 50 ppm) far greater than tralomethrin water solubility, and because of this, the information is not included in this review. The second study, conducted by Wang (1990c), utilized concentrations at or below tralomethrin water solubility. Nonlinear and linear first-order rate data from sterile samples incubated in the dark at 25 °C are presented in Table 61.

The data in Table 61 are compromised by two factors: rapid conversion of a part but not all parent tralomethrin on day zero to deltamethrin at all pH values, and incomplete transfer of tralomethrin from buffer to organic extractant. Material not extracted was excluded from further analysis. Assays at 30 d in the pH 7 buffer showed 50% of the ^{14}C in the aqueous phase was unextracted tralomethrin (8% of applied). Overall, the aqueous phases contained up to 30% of applied in the experiment. Assuming the same distribution as in the organic extract, the tralomethrin remaining in the water was calculated and then added to the organic extract to provide new values for tralomethrin amounts in each hydrolysis sample. These values were then used in nonlinear and linear plots to generate the values cited in Table 61.

2. Photolysis in Water. Deltamethrin photolysis in buffered water was studied by Bowman and Carpenter in 1987 (Table 62) in the laboratory using a xenon arc lamp at half the light intensity of sunlight, so that 1 d of continuous irradiation was equal to 1 d of natural sunlight assuming 12 hr light/dark. Dark control samples were stable, and no correction for dark reaction was required.

Photolysis was reexamined by Wang and Reynolds (1991a) who irradiated sterile buffer solutions of ^{14}C-labeled deltamethrin with light from a Heraeus Suntest CPS xenon arc lamp with UV filters. Irradiation was 12 hr light followed by 12 hr dark, and dark controls were included. The nonlinear and linear rate constants from this experiment were corrected for dark reaction and are

Table 58. Tralomethrin soil desorption partition coefficients.

Expt rating	Soil conc, ppm	Soil texture	K_{dd}	K_{ocd}	Reference
3	0.2	SL	1,730	433,000	Daly (1989)
3	0.2	S	600	240,000	Daly (1989)
3	0.2	S	1,080	432,000	Daly (1989)
3	0.2	SL	1,650	413,000	Daly (1989)
3	0.2	CL	1,290	99,200	Daly (1989)
3	0.2	L	1,640	364,000	Daly (1989)
3	0.2	L	2,600	578,000	Daly (1989)
3	0.2	CL	1,210	93,400	Daly (1989)
3	0.8	S	838	335,000	Daly (1989)
3	0.9	SL	1,950	488,000	Daly (1989)
3	0.9	S	1,500	600,000	Daly (1989)
3	0.9	CL	1,660	128,000	Daly (1989)
3	0.9	SL	3,120	780,000	Daly (1989)
3	0.9	CL	1,620	125,000	Daly (1989)
3	0.9	L	2,480	551,000	Daly (1989)
3	0.9	L	2,430	540,000	Daly (1989)
3	1	CL	2,060	158,000	Daly (1989)
3	1	L	3,100	689,000	Daly (1989)
3	1	S	2,190	876,000	Daly (1989)
3	1	L	2,220	493,000	Daly (1989)
3	1	S	1,470	588,000	Daly (1989)
3	1	SL	3,460	865,000	Daly (1989)
3	1	CL	2,190	168,000	Daly (1989)
3	1	SL	3,330	833,000	Daly (1989)
3	2	S	954	382,000	Daly (1989)
3	2	CL	2,520	194,000	Daly (1989)
3	2	SL	2,770	693,000	Daly (1989)
3	2	L	2,590	576,000	Daly (1989)
3	2	L	1,320	293,000	Daly (1989)
3	2	CL	2,270	175,000	Daly (1989)
3	2	S	1,290	516,000	Daly (1989)
3	2	SL	2,330	583,000	Daly (1989)
All ratings	Avg, <0.5		1,480	332,000	
	CV, %		39.8	52.0	
	Avg, 0.5 to <1		1,950	443,000	
	CV, %		36.3	52.1	
	Avg, 1–5		2,250	505,000	
	CV, %		32.3	50.1	
	Avg, >5				
	CV, %				
	Avg, All Conc		1,980	446,000	
	CV, %		37.5	52.5	

Table 58. (Continued).

Expt rating	Soil conc, ppm	Soil texture	K_{dd}	K_{ocd}	Reference
>4	Avg, <0.5				
	CV, %				
	Avg, 0.5 to <1				
	CV, %				
	Avg, 1–5				
	CV, %				
	Avg, >5				
	CV, %				
	Avg, All Conc				
	CV, %				

Table 59. Freundlich adsorption data for tralomethrin.

Expt rating	Soil texture	K_{df}	K_{ocf}	$1/n$	Reference
3	S	830	332,000	0.988	Daly (1989)
3	SL	3,480	870,000	1.14	Daly (1989)
3	L	197	43,800	0.738	Daly (1989)
3	CL	8,710	670,000	1.22	Daly (1989)
All ratings	Avg	3,300	479,000	1.02	
	CV, %	117.3	76.3	0.21	
>4	Avg				
	CV, %				

Table 60. Hydrolysis of deltamethrin in buffered water.

Temp, °C	pH	Nonlinear		Linear		Reference
		k^a	Half-life, d	k^a	Half-life, d	
25	5	*0.00E + 00[b]*	0	0.00E + 00	0	Smith (1990a)
25	7	*0.00E + 00[b]*	0	0.00E + 00	0	Smith (1990a)
25	9	*3.23E − 01[b]*	2.15	2.74E − 01	2.53	Smith (1990a)

[a]An entry of 0.00E + 00 means k was measured but deltamethrin is stable; a blank k or half-life value means no measurement.
[b]Representative value.

Table 61. Hydrolysis of tralomethrin in buffered water.

Temp, °C	pH	Nonlinear k	Half-life, d	Linear k	Half-life, d	Reference
25	4	5.84E − 03	119	0.00E + 00[a]	0	Wang (1990c)
25	5	2.47E − 03	281	2.44E − 03	284	Wang (1990c)
	Avg	$4.16E − 03^{b}$				
	CV, %	57.4				
25	7	$4.46E − 02^{b}$	15.5	2.34E − 02	29.6	Wang (1990c)
25	9	$1.78E − 02^{b}$	38.9	9.92E − 03	69.8	Wang (1990c)

[a]An entry of 0.00E + 00 means k was measured but tralomethrin is stable; a blank k or half-life value means no measurement.
[b]Representative value.

cited in Table 62. They reflect the results from data combined from experiments with deltamethrin labeled with ^{14}C at two different positions.

A third study by Wang (1991a) was actually a tralomethrin experiment that provided deltamethrin data as first breakdown product, in addition to tralomethrin data. Methodology and light source was the same as that used by Wang and Reynolds (1991a); dark controls were included so that kinetics could be corrected for dark reaction. A deltamethrin nonlinear first-order rate constant was calculated from the data by modeling deltamethrin formation and subsequent disappearance as a sequential first-order reaction:

$$A \rightarrow B \rightarrow C$$

where A = tralomethrin and B = deltamethrin. The nonlinear rate constant obtained from this exercise is cited in Table 62. The rate constant selected to repre-

Table 62. Photolysis of deltamethrin in water.

Natural sun	Buffer medium	Nonlinear k	Half-life, d	Linear k	Half-life, d	Reference
No	Yes	1.54E − 03	450	1.46E − 03	475	Wang and Reynolds (1991a)
No	Yes	2.03E − 02	34.1			Wang (1991a)
No	Yes	1.58E − 02	43.9	1.45E − 02	47.8	Bowman and Carpenter (1987)
	Avg	$1.25E − 02^{a}$		7.98E − 03		
	CV, %	78.1		115.5		

[a]Representative value.

Table 63. Photolysis of tralomethrin in water.

Natural sun	Buffer medium	Nonlinear		Linear		Reference
		k	Half-life, d	k	Half-life, d	
No	Yes	$2.81E-01$[a]	2.47	$9.73E-02$	7.12	Wang (1991a)
No	No					Devaux and Bolla (1984)

[a]Representative value.

sent the photolysis of deltamethrin in water is the average of the three values contained in Table 62. Two values show a moderate rate of photolysis, and a third a slow rate, and so they were averaged to reflect the totality of all results.

Tralomethrin water photolysis (Table 63) was studied by Devaux and Bolla (1984) and again by Wang (1991a). The study by Devaux and Bolla was at very high concentration (600 ppm) as a water emulsion containing emulsifier and cosolvent, and for this reason the data are not included in this review. The study by Wang (1991a) is the same one described earlier for deltamethrin modeled in a sequential first-order reaction. Dark controls were carried in the study and tralomethrin analysis at 14 and 30 d showed tralomethrin as stable with no need for dark reaction correction of the kinetic data.

3. Photolysis on Soil. Wang and Reynolds (1991b) used the same setup for the photolysis of deltamethrin on soil as that used to study its photolysis in water. Soil plates with deltamethrin added to their surfaces were brought to 75% 1/3 bar moisture, sealed in glass plates to prevent drying, and then irradiated (12 hr light/12 hr dark) with the Heraeus Suntest CPS xenon arc lamp + UV filter adjusted to be similar to New Jersey sunlight. Deltamethrin was degraded nearly as fast in the dark controls as in irradiated samples; the data cited in Table 64 are corrected for dark reaction.

An early study of tralomethrin photolysis on soil by Warren (1984) provided

Table 64. Photolysis of deltamethrin on soil.

Natural sun	Moisture state	Nonlinear		Linear		Reference
		k	Half-life, d	k	Half-life, d	
No	Wet	$2.00E-02$[a]	34.7	$1.10E-02$	63	Wang and Reynolds (1991b)

[a]Representative value.

Table 65. Photolysis of tralomethrin on soil.

Natural sun	Moisture state	Nonlinear		Linear		Reference
		k	Half-life, d	k	Half-life, d	
No	Air dry	$1.79E-01$[a]	3.87	$1.31E-02$	52.9	Wang (1990a)
No	Air dry					Warren (1984)

[a]Representative value.

inconclusive data for tralomethrin (Table 65) because of rapid, random conversion of tralomethrin to deltamethrin. Its data are assumed to be suited better for assessment of the combination of tralomethrin + deltamethrin, which is discussed later. Tralomethrin was studied again in 1990 by Wang (1990a) on air-dry soil irradiated with a Heraeus Sun Test CPS xenon arc lamp for alternating 12-hr periods of light and dark. Dark controls were included, and data in Table 65 are corrected for dark reaction. In this experiment tralomethrin degraded rapidly to deltamethrin, which was separated analytically from tralomethrin. However, the rapid loss of tralomethrin from the soil surface and its impact on subsequent exposure to light caused the kinetics to deviate from first order. For that reason, it was decided to use only data from the first 7 d of irradiation for nonlinear regression analysis; all data through 30 d of irradiation were used for the linear analysis.

C. Biotic Chemical Properties

1. Aerobic Soil Degradation. Aerobic degradation of deltamethrin has been measured in several soils by Kaufman et al. (1990a,b) and by Wang (1990b,c) at concentrations less than 0.5 ppm applied to soil biometer flasks. Chemical was added in 100 μL of ethanol or acetonitrile per 50 g soil, quantities small enough to have little impact on the kinetic outcome. Results from the nonlinear and linear regression of datasets from each experiment are given in Table 66.

Table 67 summarizes the tralomethrin aerobic soil data. Wang (1990b) applied tralomethrin in 100 μL acetonitrile to 50 g soil in biometer flasks incubated at 25 °C and 75% 1/3 bar moisture in the dark up to 182 d. Analytical methodology separated tralomethrin from deltamethrin (the first breakdown product) and allowed evaluation of the rate with which tralomethrin degrades in soil.

2. Anaerobic Soil Degradation. Wang (1990d) added tralomethrin in 100 μL acetonitrile to 50 g soil in biometer flasks incubated for the first 3 d aerobically at 75% 1/3 bar moisture and then anaerobically thereafter by waterlogging and purging samples with nitrogen gas. During the 3 d of aerobic incubation most of the tralomethrin was converted to deltamethrin, making this study largely an

Table 66. Rate of deltamethrin degradation in aerobic soil.

Expt rating	Temp, °C	Soil conc, ppm	Soil texture	Nonlinear		Linear		Reference
				k	Half-life, d	k	Half-life, d	
8	25	0.01	SiL	2.44E − 02	28.4	1.44E − 02	48.1	Kaufman et al. (1990b
8	25	0.01	FSL	3.26E − 02	21.3	1.29E − 02	53.7	Kaufman et al. (1990b
8	25	0.1	SiL	2.79E − 02	24.8	1.42E − 02	48.8	Kaufman et al. (1990b
8	25	0.1	FSL	2.60E − 02	26.7	1.32E − 02	52.5	Kaufman et al. (1990b
7	25	0.1	FSL	5.09E − 02	13.6	3.51E − 02	19.7	Kaufman et al. (1990b
8	25	0.2	SL	1.80E − 02	38.5			Wang (1990b)
8	25	0.2	SL	3.02E − 02	23	1.52E − 02	45.6	Wang (1991c)
8	25	0.2	SL	3.64E − 02	19	1.91E − 02	36.3	Wang (1991c)
7	25	0.2	FSL	3.08E − 02	22.5	2.82E − 02	24.6	Kaufman et al. (1990a
All ratings		Avg, <0.5			24.2		41.2	
		CV, %			28.5		31.4	
		Avg + CL			27.4		47.5	
		Avg, 0.5–1.5						
		CV, %						
		Avg + CL						
		Avg, 1.5–10						
		CV, %						
		Avg + CL						
		Avg, >10						
		CV, %						
		Avg + CL						
		Avg, All Conc			24.2		41.2	
		CV, %			28.5		31.4	
		Avg + CL			27.4		47.5	
>4		Avg, All Conc			24.2[a]		41.2	
		CV, %			28.5		31.4	
		Avg + CL			27.4[a]		47.5	

[a]Representative value.

anaerobic experiment with deltamethrin. The data cited in Table 68 are anaerobic data for deltamethrin.

Anaerobic degradation of deltamethrin was reexamined by Wang (1991b) who added deltamethrin in 100 μL acetonitrile to 50 g soil in biometer flasks, which were incubated aerobically for 15 d (1 half-life) and then made anaerobic by waterlogging and purging samples with nitrogen gas. Kinetic data from anaerobic incubation to 90 d are cited in Table 68.

Results for tralomethrin from the anaerobic study with tralomethrin by Wang (1990d) as discussed for deltamethrin are reported in Table 69. The rapid loss of tralomethrin during the aerobic first 3 d of the study did not provide adequate

Table 67. Rate of tralomethrin degradation in aerobic soil.

Expt rating	Temp, °C	Soil conc, ppm	Soil texture	Nonlinear		Linear		Reference
				k	Half-life, d	k	Half-life, d	
8	25	0.2	SL	2.13E – 01	3.25	2.86E – 02	24.2	Wang (1990b)
All ratings		Avg, <0.5 CV, %			3.25		24.2	
		Avg + CL			9.75		72.6	
		Avg, 0.5–1.5 CV, % Avg + CL						
		Avg, 1.5–10 CV, % Avg + CL						
		Avg, >10 CV, % Avg + CL						
		Avg, All Conc CV, %			3.25		24.2	
		Avg + CL			9.75		72.6	
>4		Avg, All Conc CV, %			3.25[a]		24.2	
		Avg + CL			9.75[a]		72.6	

[a]Representative value.

data for regression analysis, and a linear rate constant was calculated in the following manner. At 3 d 29.3% of the original tralomethrin remained, but by the next sampling period at 30 d all tralomethrin was gone. Therefore, it was assumed that disappearance of chemical was governed by first order, and that 6 half-lives had passed, to bring the material remaining to 0.5% of applied. Six half-lives within a 30-d period is equivalent to the 5-d half-life reported in Table 69.

3. Aerobic Aquatic Degradation. Deltamethrin degradation in aerobic aquatic systems was studied by Muttzall (1993), who applied deltamethrin in 10 μL DMSO to 200 mL natural water + 20 g sediment in biometer flasks. Samples were incubated on a rotary shaker in the dark at 20 °C for up to 84 d. Nonlinear and linear regression analyses of the deltamethrin remaining in the samples are

Table 68. Deltamethrin anaerobic degradation generated by anaerobic soil experimental design.

Expt rating	Temp, °C	Soil conc, ppm	Soil texture	Nonlinear		Linear		Reference
				k	Half-life, d	k	Half-life, d	
6	25	0.2	SL	2.63E – 02	26.4	2.05E – 02	33.8	Wang (1991b)
6	25	0.2	SL	2.20E – 02	31.5	1.99E – 02	34.8	Wang (1990d)
All ratings		Avg, <1.5			28.9		34.3	
		CV, %			12.5		2.1	
		Avg + CL			36.8		35.8	
		Avg, >1.5						
		CV, %						
		Avg + CL						
		Avg, All Conc			28.9		34.3	
		CV, %			12.5		2.1	
		Avg + CL			36.8		35.8	
>4		Avg, All Conc			28.9[a]		34.3	
		CV, %			12.5		2.1	
		Avg + CL			36.8[a]		35.8	

[a]Representative value.

summarized in Table 70. No information was available for aerobic aquatic degradation of tralomethrin.

4. Anaerobic Aquatic Degradation. There were no anaerobic experiments for either deltamethrin or tralomethrin utilizing the anaerobic aquatic design.

D. Tralo + Deltamethrin

Some tralomethrin experiments used tralomethrin as starting material, but because analytical methodology was insufficient to distinguish tralomethrin from deltamethrin, results were reported as the sum of the two chemicals. Thus, these studies do not compare directly with deltamethrin or tralomethrin experiments, and they are kept separate in this review. They do not appear in the first section that summarizes the physical and chemical properties of the pyrethroids.

1. Photolysis on Soil. One such study (Table 71) was soil photolysis of tralomethrin, carried out by Warren (1984). Tralomethrin in 45 mL dichloromethane was applied to 50 g air-dry soil at 10 ppm, and after solvent evaporation was mixed thoroughly into the soil. One-gram portions were placed in uncapped scintillation vials that were irradiated continuously by light from an Ace-Ha-

Table 69. Tralomethrin anaerobic degradation generated by anaerobic soil experimental design.

Expt rating	Temp, °C	Soil conc, ppm	Soil texture	Nonlinear		Linear		Reference
				k	Half-life, d	k	Half-life, d	
6	25	0.2	SL			1.39E – 01	5	Wang (1990d)
All ratings		Avg, <1.5 CV, %					5	
		Avg + CL					15	
		Avg, >1.5 CV, % Avg + CL						
		Avg, All Conc CV, %					5	
		Avg + CL					15	
>4		Avg, All Conc CV, %					5[a]	
		Avg + CL					15[a]	

[a]Representative value.

novia high pressure Hg vapor lamp (light intensity not characterized). Tralomethrin was not separated analytically from deltamethrin, the first breakdown product, and so the results in Table 71 are in terms of tralo + deltamethrin. Because irradiation was continuous, first-order rate constants in Table 71 are corrected by a factor of 2 (light reaction kinetics only). Data are not rated highly in level of confidence because of the drench application of high tralomethrin rates followed by subsequent mixing of the soil before irradiation, causing considerable shielding of chemical from exposure to light. Poor linear and nonlinear fit to first order was observed for the tralo + deltamethrin dataset over the full 30-d period.

2. Aerobic Soil Degradation.   Aerobic soil experiments by Kaufman et al. (date unknown b) began with tralomethrin but used analytical methods that did not distinguish tralomethrin from deltamethrin. Tralomethrin was applied in 100 μL ethanol to 50 g soil in biometer flasks and samples incubated at 25 °C and 75% 1/3 bar moisture up to 128 d. Results from nonlinear and linear first-order kinetic analysis of the data are presented in Table 72.

3. Anaerobic Soil Degradation.   Kaufman et al. (date unknown c) examined the anaerobic soil degradation of tralomethrin but did not use analytical method-

Table 70. Deltamethrin degradation in aerobic aquatic systems.

Sample name	Sediment conc, ppm	Nonlinear		Linear		Reference
		k	Half-life, d	k	Half-life, d	
TNO Ditch	1.4	1.58E − 02	43.9	1.30E − 02	53.3	Muttzall (1993)
Kromme Rijn	1.4	6.02E − 03	115	5.98E − 03	116	Muttzall (1993)
	Avg		79.5[a]		84.7	
	CV, %		63.3		52.4	
	Avg + CL		189[a]		181	

[a]Representative value.

Table 71. Photolysis of tralo + deltamethrin on soil.

Natural sun	Moisture state	Nonlinear		Linear		Reference
		k	Half-life, d	k	Half-life, d	
No	Air dry	$1.34E-03$[a]	517	$8.70E-04$	797	Warren (1984)

[a]Representative value.

ology capable of distinguishing tralomethrin from deltamethrin. Tralomethrin in 100 µL of ethanol was added to 50 g soil in soil biometer flasks and then incubated aerobically for 32 d before being made anaerobic by flooding and purging with nitrogen gas. Kinetic analysis of the data is summarized in Table 73.

4. Anaerobic Aquatic Degradation.   While studying anaerobic decomposition of tralomethrin, Kaufman et al. (date unknown c) used an anaerobic aquatic design in addition to the anaerobic soil experimental design already cited. Samples were either treated with tralomethrin at time of flooding or 32 d after flooding, and the incubation headspace was purged with nitrogen gas. Some samples also included amendment with hay or soybean to encourage the formation of anaerobicity. Results from samples incubated to 128 d (length of experiment) are summarized in Table 74.

VII. Esfenvalerate

Esfenvalerate is the "S" enantiomer of fenvalerate, which is included under the "Esfenvalerate" heading. Esfenvalerate is [S-(R^*,R^*)]-cyano(3-phenoxyphenyl)-methyl 4-chloro-2-(1-methylethyl)benzeneacetate [IUPAC = (S)-α-cyano-3-phenoxybenzyl (S)-2-(4-chlorophenyl)-3-methylbutyrate]. Typical composition of technical esfenvalerate is $SR = 7\%$, $RS = 7\%$, $SS = 85\%$, and $RR = 1\%$. Other names referring to esfenvalerate are DPX-GB800, Asana®, S-1844, S-5602 Alpha, WL 43775, and SD 43775. CAS RN is 66230-04-4; official code is OMS 3023 (Tomlin 1994).

Technical fenvalerate is the mixture of all four enantiomers in the ratio: $SR = 27\%$, $RS = 27\%$, $SS = 23\%$, and $RR = 23\%$. Its IUPAC name is (RS)-α-cyano-3-phenoxybenzyl (RS)-2-(4-chlorophenyl)-3-methylbutyrate. Fenvalerate has been referred to S-5602, WL 43775, and Pydrin®; its CAS RN is 51630-58-1 and the official code is OMS 2000.

A. Physical Properties

1. Vapor Pressure.   Original references citing data for the vapor pressure of esfenvalerate were not available for this review; vapor pressure information presented in Table 75 was extracted from Tomlin (1994).

Table 72. Tralo + deltamethrin degradation in aerobic soil.

Expt rating	Temp, °C	Soil conc, ppm	Soil texture	Nonlinear		Linear		Reference
				k	Half-life, d	k	Half-life, d	
8	25	0.012	SiL	9.37E – 03	74	1.04E – 02	66.6	Kaufman et al. (date unknown ▶
8	25	0.012	SiL	2.67E – 02	26	1.77E – 02	39.2	Kaufman et al. (date unknown ▶
8	25	0.012	FSL	1.41E – 02	49.2	1.37E – 02	50.6	Kaufman et al. (date unknown ▶
8	25	0.049	SiL	8.12E – 03	85.4	9.41E – 03	73.7	Kaufman et al. (date unknown ▶
8	25	0.049	SiL	2.34E – 02	29.6	1.66E – 02	41.8	Kaufman et al. (date unknown ▶
8	25	0.049	FSL	1.16E – 02	59.8	1.29E – 02	53.7	Kaufman et al. (date unknown ▶
8	25	0.1	SiL	1.52E – 02	45.6	1.24E – 02	55.9	Kaufman et al. (date unknown ▶
All ratings		Avg, <0.5			52.8		54.5	
		CV, %			41.5		22.8	
		Avg + CL			64.7		61.3	
		Avg, 0.5–1.5						
		CV, %						
		Avg + CL						
		Avg, 1.5–10						
		CV, %						
		Avg + CL						
		Avg, >10						
		CV, %						
		Avg + CL						
		Avg, All Conc			52.8		54.5	
		CV, %			41.5		22.8	
		Avg + CL			64.7		61.3	
>4		Avg, All Conc			52.8[a]		54.5	
		CV, %			41.5		22.8	
		Avg + CL			64.7[a]		61.3	

[a]Representative value.

2. *Water Solubility.*   Original references for the water solubility of esfenvalerate were not available. Data cited in Table 76 were taken from Tomlin (1994).

3. *Henry's Law Constant.*   Calculation of K_h for esfenvalerate by Eq. 1 and assuming a molecular weight of 419.9, vapor pressure of 1.5×10^{-9} mm Hg, and water solubility of 6.0×10^{-3} ppm yields a K_h value of 1.4×10^{-7} atm m^3 mol^{-1}.

Table 73. Tralo + deltamethrin anaerobic degradation generated by anaerobic soil design.

Expt rating	Temp, °C	Soil conc, ppm	Soil texture	Nonlinear		Linear		Reference
				k	Half-life, d	k	Half-life, d	
7	25	0.012	SiL	2.86E – 03	242	2.57E – 03	270	Kaufman et al. (unknown date c)
7	25	0.012	SiL	1.47E – 02	47.2	6.65E – 03	104	Kaufman et al. (unknown date c)
7	25	0.012	SiL	2.10E – 03	330	2.11E – 03	329	Kaufman et al. (unknown date c)
7	25	0.012	FSL	5.27E – 03	132	7.21E – 03	96.1	Kaufman et al. (unknown date c)
All ratings		Avg, <1.5			188		200	
		CV, %			66.0		58.9	
		Avg + CL			289		296	
		Avg, >1.5						
		CV, %						
		Avg + CL						
		Avg, All Conc			188		200	
		CV, %			66.0		58.9	
		Avg + CL			289		296	
>4		Avg, All Conc			188[a]		200	
		CV, %			66.0		58.9	
		Avg + CL			289[a]		296	

[a]Representative value.

4. Octanol–Water Partition Coefficient.   No original references were available for review. K_{ow} values cited in Table 77 were extracted from Tomlin (1994). Two values are reported; the first is for esfenvalerate and the second for fenvalerate. The two values were averaged in Table 77, and the average was selected to represent esfenvalerate. K_{ow} calculated from the molecular structure of esfenvalerate is also provided in the table as CLog P; Log P calculated from structure is higher than the values extracted from Tomlin (1994).

5. Bioconcentration Factor.   Carp were exposed continuously to fenvalerate over a 28-d period by Lee (1990). Water and fish were analyzed periodically for total ^{14}C and fenvalerate levels so that BCF could be calculated on the basis of total ^{14}C and fenvalerate concentrations. Data cited in Table 78 are the value for BCF at day 28, based on measured fenvalerate concentrations in whole fish at 28 d and average concentrations in water over the 28-d period.

Table 74. Tralo + deltamethrin degradation generated by an anaerobic aquatic experimental design.

Expt rating	Temp, °C	Soil conc, ppm	Soil texture	Nonlinear		Linear		Reference
				k	Half-life, d	k	Half-life, d	
7	25	0.012	FSL	9.14E – 03	75.8	1.22E – 02	56.8	Kaufman et al. (date unknown c)
7	25	0.012	FSL	6.70E – 03	103	6.43E – 03	108	Kaufman et al. (date unknown c)
7	25	0.012	FSL	1.23E – 02	56.4	1.05E – 02	66	Kaufman et al. (date unknown c)
7	25	0.012	FSL	7.05E – 03	98.3	5.68E – 03	122	Kaufman et al. (date unknown c)
7	25	0.012	FSL	9.11E – 03	73.1	1.30E – 02	53.3	Kaufman et al. (date unknown c)
7	25	0.012	FSL	6.44E – 03	108	6.36E – 03	109	Kaufman et al. (date unknown c)
All ratings		Avg, <1.5			85.8		85.9	
		CV, %			23.7		35.4	
		Avg + CL			98.0		104	
		Avg, >1.5						
		CV, %						
		Avg + CL						
		Avg, All Conc			85.8		85.9	
		CV, %			23.7		35.4	
		Avg + CL			98.0		104	
>4		Avg, All Conc			85.8[a]		85.9	
		CV, %			23.7		35.4	
		Avg + CL			98.0[a]		104	

[a]Representative value.

Table 75. Esfenvalerate vapor pressure.

Temperature, °C	Best value	VP, mm Hg	Source
20	No	1.44E – 07	Tomlin (1994)
25	No	1.50E – 09[a]	Tomlin (1994)

[a]Representative value.

Table 76. Water solubility of esfenvalerate.

Temperature, °C	Column saturation	WS, ppm	Reference
25	No	1.00E − 02	Tomlin (1994)
25	No	2.00E − 03	Tomlin (1994)
	Avg	6.00E − 03[a]	
	CV, %	94.3	

[a]Representative value.

Table 77. Esfenvalerate octanol–water partition coefficient.

Best values	K_{ow}	Log P	Clog P	Reference
No	1.66E + 06	6.22		Tomlin (1994)
No	1.02E + 05	5.01		Tomlin (1994)
Avg	8.81e + 02	5.62[a]	6.8[b]	
CV, %	125	15.2		

[a]Representative value.
[b]Calculated from molecular structure by Briggs (1999).

6. Soil Sorption.   Soil sorption data for esfenvalerate or fenvalerate were not available.

B. Abiotic Chemical Properties

1. Hydrolysis.   Hydrolysis studies with esfenvalerate are summarized in Table 79. Lee (1988) added esfenvalerate to sterile pH 5, 7, or 9 buffers and incubated them in the dark up to 30 d. Analysis indicated no breakdown of esfenvalerate at any pH values; esfenvalerate was shown to be stable to hydrolysis in water.

2. Photolysis in Water.   Mikami et al. (1980) added fenvalerate with Tween 85 emulsifier to water in quartz flasks that were exposed to Japanese summer sunlight up to 42 d. Data from only two sampling periods (not zero time) were

Table 78. Esfenvalerate bioconcentration factor based on fish whole-body analysis.

Type of experiment	Best values	BCF	Reference
Flow-through	Yes	2,390[a]	Lee (1990)

[a]Representative value.

Table 79. Hydrolysis of esfenvalerate in buffered water.

Temp, °C	pH	Nonlinear		Linear		
		k^a	Half-life, d	k	Half-life, d	Reference
25	5	$0.00E + 00^b$	0	$0.00E + 00$	0	Lee (1988)
25	7	$0.00E + 00^b$	0	$0.00E + 00$	0	Lee (1988)
25	9	$0.00E + 00^b$	0	$0.00E + 00$	0	Lee (1988)

[a]An entry of 0.00E + 00 means k was measured but esfenvalerate is stable; a blank k or half-life value means no measurement.
[b]Representative value.

given in the report, and so a first-order rate constant was estimated by the use of the first-order point equation described in Eq. 4.

A second esfenvalerate study by Stevenson (1987) was carried out in buffer plus a small amount of acetonitrile, using continuous irradiation from a Heraeus "Suntest" xenon lamp with filters that simulated two days of June sunlight in Delaware, assuming 12-hr light/dark periods. Time intervals of irradiated samples (not controls) were lengthened by a factor of 2 to compensate, and kinetics are corrected for dark reaction with data from the dark control samples. Because this study used water without emulsifier, it was chosen as the source of data for model input, even though the experiment by Mikami et al. utilized natural sunlight. Data are summarized in Table 80.

3. *Photolysis on Soil.* Photolysis of esfenvalerate on soil surfaces has been studied by Mikami et al. (1980), by Ehman and Ingamells (1981), and by Castle et al. (1990). All utilized natural sunlight as source of irradiation and all except Castle et al. performed their experiments with air-dry soil.

Mikami et al. exposed air-dry soil TLC plates to Japanese fall sunlight for 3 or 10 d (no zero day analysis); first-order rate data (Table 81) corrected for dark control reaction were calculated from the first-order point equation (Eq. 4). Ehman and Ingamells (1981) irradiated soil TLC plates in outdoor California sun-

Table 80. Photolysis of esfenvalerate in water.

Natural sun	Buffer medium	Nonlinear		Linear		
		k	Half-life, d	k	Half-life, d	Reference
No	Yes	$4.02E - 02^a$	17.2	$3.84E - 02$	18.1	Stevenson (1987)
Yes	No			$9.89E - 02$	7.01	Mikami et al. (1980)

[a]Representative value.

light for up to 42 d; Castle et al. (1990) used thin layers of wet soil in petri dishes to expose esfenvalerate to California winter sunlight up to 30 d. All studies included dark controls, and kinetic data from them are corrected for dark reaction. Results of the kinetic analyses of the three studies are summarized in Table 81. The study conducted with wet soil had no measurable photolysis of esfenvalerate; the two studies with air-dry soil showed fairly rapid photolysis. For selection of input for modeling purposes, it was decided to average all values in Table 81 and select the mean value as representation of esfenvalerate on soil surfaces.

C. Biotic Chemical Properties

1. Aerobic Soil Degradation. Table 82 shows that several studies of the aerobic degradation of esfenvalerate have been conducted in a variety of soils and over a range of application rates. All studies applied chemical to soils in an appropriate organic solvent and then incubated the treated soils aerobically at moisture values near optimum. In some experiments (Lee 1979; Lee and Stackhouse 1979; Potter and Arnold 1980; Lee et al. 1985; Gaddamidi and Bookhart 1992), volumes of organic solvent and application rates were high enough to possibly impact the kinetics of the degradation process, causing less certainty in the level of confidence for data from those studies. These studies were assigned lesser experimental ratings and were separated from experiments carried out with lower application rates and lower volumes of organic solvent.

2. Anaerobic Soil Degradation. Studies on the degradation of esfenvalerate in anaerobic soil experimental designs have been reported by Lee (1985) and Gaddamidi and Bookhart (1992) (Table 83). Lee added fenvalerate in methanol

Table 81. Photolysis of esfenvalerate on soil.

Natural sun	Moisture state	Nonlinear		Linear		Reference
		k	Half-life, d	k	Half-life, d	
Yes	75% 1/3 bar	0.00E + 00	0	0.00E + 00	0	Castle et al. (1990)
Yes	Air dry			4.13E − 02	16.8	Mikami et al. (1980)
Yes	Air dry			4.02E − 02	17.2	Mikami et al. (1980)
Yes	Air dry	1.03E − 01	6.73	4.80E − 02	14.4	Ehman and Ingamells (1981)
Yes	Air dry	1.05E − 01	6.6	3.40E − 02	20.4	Ehman and Ingamells (1981)
	Avg	6.93E − 02[a]				
	CV, %	86.6				

[a]Representative value.

Table 82. Esfenvalerate degradation in aerobic soil.

Expt rating	Temp, °C	Soil conc, ppm	Soil texture	Nonlinear k	Nonlinear Half-life, d	Linear k	Linear Half-life, d	Reference
6	20	0.51	LS	3.41E – 02	20.3	1.95E – 02	35.5	Itoh et al. (1995)
6	20	0.51	LS	2.89E – 02	24	1.72E – 02	40.3	Itoh et al. (1995)
6	20	0.51	Sand	1.90E – 02	36.5	1.63E – 02	42.5	Itoh et al. (1995)
6	20	0.51	Sand	1.32E – 02	52.5	1.17E – 02	59.2	Itoh et al. (1995)
6	20	0.51	SL	1.71E – 02	40.5	1.41E – 02	49.2	Itoh et al. (1995)
6	25	1	SCL			6.19E – 03	112	Mikami et al. (1984a)
5	25	1	Light Clay			4.62E – 02	15	Ohkawa et al. (1978)
5	25	1	SCL			7.70E – 03	90	Ohkawa et al. (1978)
6	25	1	Light Clay			3.30E – 02	21	Mikami et al. (1984a)
5	16	1	CL	1.18E – 02	58.7	1.37E – 02	50.6	Williams and Brown (1979)
5	16	1	SiL	1.29E – 02	53.7	1.30E – 02	53.3	Williams and Brown (1979)
6	20	1.52	LS	1.85E – 02	37.5	1.45E – 02	47.8	Itoh et al. (1995)
6	20	1.52	LS	2.95E – 02	23.5	1.98E – 02	35	Itoh et al. (1995)
4	25	2.5	SL	2.07E – 02	33.5	1.54E – 02	45	Gaddamidi and Bookhart (1992)
3	25	5	SiCL	5.69E – 03	122	5.37E – 03	129	Lee (1979)
4	22	5	SL	1.26E – 02	55	1.23E – 02	56.4	Lee and Stackhouse (1979)
2	25	5	SiL	9.64E – 03	71.9	9.40E – 03	73.7	Lee et al. (1985)
2	25	5	SiL	1.18E – 03	587	1.27E – 03	546	Potter and Arnold (1980)
2	25	5	SiCL	3.36E – 03	206	3.42E – 03	203	Potter and Arnold (1980)
2	25	5	SL	4.46E – 03	155	4.19E – 03	165	Potter and Arnold (1980)
3	25	5	SiL	2.43E – 03	285	2.61E – 03	266	Lee (1979)
3	25	5	SL	6.89E – 03	101	5.15E – 03	135	Lee (1979)
2	25	5	SL	4.08E – 03	170	3.84E – 03	181	Potter and Arnold (1980)
2	25	20	SiL	5.74E – 03	121	5.95E – 03	116	Lee et al. (1985)

All ratings	Avg, <0.5							
	CV, %							
	Avg + CL							
	Avg, 0.5–1.5				40.9		51.7	
	CV, %				36.6		54.5	
	Avg + CL				49.0		63.3	

Table 82. (Continued).

Expt rating	Temp, °C	Soil conc, ppm	Soil texture	Nonlinear		Linear		Reference
				k	Half-life, d	k	Half-life, d	
		Avg, 1.5–10			154		157	
		CV, %			102.4		90.9	
		Avg + CL			216		213	
		Avg, >10			121		116	
		CV, %						
		Avg + CL			363		348	
		Avg, All Conc			113		107	
		CV, %			117.1		105.9	
		Avg + CL			152		137	
>4		Avg, All Conc			38.6[a]		50.1	
		CV, %			36.8		52.2	
		Avg + CL			45.2[a]		59.9	

[a]Representative value.

(250 μL/50 g) to soil at a fairly high rate of 5 ppm and incubated treated samples aerobically for the first 30 d and then anaerobically for an additional 30 or 60 d by the establishment of a water layer over the soil and change of incubation atmosphere to nitrogen.

Gaddamidi and Bookhart added esfenvalerate in 536 μL acetone to 50 g soil at a lower concentration of 2.5 ppm and incubated the samples aerobically for 30 d. They then made them anaerobic for up to 60 d by flooding them with water, purging the atmospheres with nitrogen gas, and sealing the incubation containers to prevent reintroduction of atmospheric oxygen. Nonlinear and linear kinetic results from these studies are summarized in Table 83.

3. Aerobic Aquatic Degradation. Lewis (1995) examined the fate of esfenvalerate in aerobic aquatic systems (Table 84). Esfenvalerate in 48 μL acetonitrile was added dropwise to the surface of multiple samples of each of two water–sediment systems in glass cylinders, which were then incubated at 10 °C in the dark for up to 100 d. Kinetic analysis of the data is summarized in Table 84.

4. Anaerobic Aquatic Degradation. No anaerobic degradation data were available from experiments utilizing an anaerobic aquatic experimental design.

VIII. Fenpropathrin

The fenpropathrin C.A. name is cyano(3-phenoxyphenyl)methyl 2,2,3,3-tetramethylcyclopropanecarboxylate [IUPAC = (*RS*)-α-cyano-3-phenoxybenzyl 2,2,

Table 83. Esfenvalerate degradation generated by an anaerobic soil experimental design.

Expt rating	Temp, °C	Soil conc, ppm	Soil texture	Nonlinear		Linear		Reference
				k	Half-life, d	k	Half-life, d	
5	25	2.5	SL	7.67E − 03	90.4	6.64E − 03	104	Gaddamidi and Bookhart (1992)
4	23	5	SL	3.44E − 03	201	3.42E − 03	203	Lee (1985)
All ratings		Avg, <1.5 CV, % Avg + CL						
		Avg, >1.5 CV, % Avg + CL			146 53.7 316		154 45.6 306	
		Avg, All Conc CV, % Avg + CL			146 53.7 316		154 45.6 306	
>4		Avg, All Conc CV, % Avg + CL			90.4[a] 271[a]		104 312	

[a]Representative value.

3,3-tetramethylcyclopropanecarboxylate]. Its molecular weight is 349.4 and the CAS RN is 64257-84-7. Other names referring to this pyrethroid are S-3206 and WL 41706. Official code is OMS 1999.

A. Physical Properties

1. Vapor Pressure.   Fenpropathrin vapor pressure has been measured by Lorence (1991), using the gas saturation method and temperatures of 25°, 35°, 40°, and 45 °C so that extrapolation of vapor pressure to lower temperature is unnecessary. The value cited in Table 85 is the mean of four measurements at 25 °C; coefficient of variation was 32.5%.

2. Water Solubility.   Fenpropathrin water solubility (Table 86) has been measured by Saito and Itoh (1992) and more recently by Lorence in 1996. Saito and Itoh used the column saturation technique to generate saturated solutions at 25 °C; the value cited from this work is the mean of 10 measurements at two different flow rates.

   Lorence (1996) saturated water by coating fenpropathrin on the walls of the glass container followed by equilibration with water stirred gently with a Teflon

Table 84. Esfenvalerate degradation in aerobic aquatic systems.

Sample name	Sediment conc, ppm	Nonlinear		Linear		Reference
		k	Half-life, d	k	Half-life, d	
Mill Stream Pond	0.13	1.06E − 02	65.4	1.10E − 02	63	Lewis (1995)
Site B	0.13	8.75E − 03	79.2	9.49E − 03	73	Lewis (1995)
	Avg		72.3[a]		68.0	
	CV, %		13.5		10.4	
	Avg + CL		93.5[a]		83.4	

[a]Representative value.

Table 85. Fenpropathrin vapor pressure.

Temperature, °C	Best value	VP, mm Hg	Source
25	Yes	$1.39E-08$[a]	Lorence (1991)

[a]Representative value.

stir bar up to 11 d at 25 °C. Samples underwent high-speed centrifugation to minimize suspension artifacts before analysis by liquid–liquid extraction and gas chromatographic analysis. Data cited in Table 86 are the mean of six measurements. Although the two citations in Table 86 are similar and because special precaution was taken by Lorence to minimize suspension effects, the more recent value by Lorence was selected as the representative value.

3. Henry's Law Constant. Calculation of K_h by Eq. 1, assuming a molecular weight of 349.4, a vapor pressure of 1.39×10^{-8} mm Hg, and water solubility of 1.03×10^{-2} yields $K_h = 6.2 \times 10^{-7}$ atm m^3 mol^{-1}.

4. Octanol–Water Partition Coefficient. K_{ow} for fenpropathrin (Table 87) was measured in 1983 (Valent) by equilibration of water with octanol containing fenpropathrin. Samples were shaken to attain equilibrium, centrifuged, and the phases analyzed for fenpropathrin. The value cited in Table 87 is the mean of four measurements having a Log P standard deviation of 0.2. CLog P calculated from the molecular structure of fenpropathrin by Briggs (1999), included in the table, is noted to be slightly less than the measured value.

5. Bioconcentration Factor. Bioconcentration of fenpropathrin in fish has been studied by Takimoto et al. (1985), by Forbis (1985), and again by Cheng (1986). The Takimoto study exposed carp continuously in glass aquaria up to 14 d and determined the ratio of total ^{14}C in fish (whole-body basis) to that in the exposure water. BCF cited in Table 88 was derived from the mean values for whole-body fish and water from days 3 to 14.

Forbis (1985) carried out a second study with bluegill exposed in a proportional diluter system for up to 28 d, monitoring fish and water for total ^{14}C. Analyses of the samples generated by Forbis were reported later by Cheng

Table 86. Water solubility of fenpropathrin.

Temperature, °C	Column saturation	WS, ppm	Reference
25	No	$1.03E-02$[a]	Lorence (1996)
25	Yes	$3.63E-02$	Saito and Itoh (1992)

[a]Representative value.

Table 87. Fenpropathrin octanol–water partition coefficient.

Best values	K_{ow}	Log P	CLog P	Reference
				Valent
Yes	1.00E + 06	6.00[a]	5.7[b]	(1983)

[a]Representative value.
[b]Calculated from molecular structure by Briggs (1999).

(1986), who included identification of ^{14}C material, The value cited in Table 88 is thus based on fenpropathrin concentrations in water and fish. Two ^{14}C labels were used, providing two datasets for calculation of BCF in this experiment; the value given in the table for Cheng is the 21-d mean from the two datasets. The representative value is the average of the two values from the Takimoto et al. and Forbis/Cheng studies.

6. Soil Sorption. Fenpropathrin sorption to and desorption from soil were measured by Lee (1992), who applied ^{14}C-labeled fenpropathrin at several concentrations in a small quantity of acetonitrile to soil in glass tubes containing dilute calcium nitrate. The solutions were equilibrated with the soil by shaking the tubes for 24 hr. K_d was determined by the difference method, which depends on measurement of ^{14}C in the supernatant after equilibration. The authors assumed that any loss of ^{14}C after equilibration is the equivalent of sorbed material. Desorption was measured by removal of supernatant, replacement with an equal amount of fresh supernatant, reequilibration, and remeasurement of ^{14}C in supernatant. Soil was combusted to obtain a direct measure of total ^{14}C in the soil solids. No identification of the ^{14}C in the water or soil was performed except for one sample of supernatant from the desorption phase. Thin-layer chromatography (TLC) analysis of this sample indicated ^{14}C was mainly polar material other than fenpropathrin, suggesting that K_d values from this study underestimated sorption and desorption of fenpropathrin.

In supplemental work, Shelby (1996) reevaluated Lee's results and recalculated the sorption and desorption values, which are presented in Tables 89, 90,

Table 88. Fenpropathrin bioconcentration factor based on fish whole-body analysis.

Type of experiment	Best values	BCF	Reference
Flow-through	Yes	355	Takimoto et al. (1985)
Flow-through	No		Forbis (1985)
Flow-through	Yes	362	Cheng (1986)
	Avg	359[a]	
	CV, %	1.4	

[a]Representative value.

Table 89. Fenpropathrin soil adsorption partition coefficients.

Expt rating	Soil conc, ppm	Soil texture	K_d	K_{oc}	Reference
4	0.0032	L	60.4	13,100	Shelby (1996)
4	0.0033	L	66	14,300	Shelby (1996)
4	0.0034	L	68	14,800	Shelby (1996)
4	0.0035	CL	152	13,300	Shelby (1996)
4	0.0037	SiL	925	97,400	Shelby (1996)
4	0.0037	CL	247	21,700	Shelby (1996)
4	0.0038	SiL	1,900	200,000	Shelby (1996)
4	0.0038	CL	200	17,500	Shelby (1996)
4	0.004		2,000	88,500	Shelby (1996)
4	0.0047	SL	276	34,500	Shelby (1996)
4	0.0068	L	71.6	15,600	Shelby (1996)
4	0.0068	L	53.5	11,600	Shelby (1996)
4	0.0071	L	81.6	17,700	Shelby (1996)
4	0.0073	CL	348	30,500	Shelby (1996)
4	0.0073	CL	270	23,700	Shelby (1996)
4	0.0074	L	206	44,800	Shelby (1996)
4	0.0075	CL	300	26,300	Shelby (1996)
4	0.0076	L	211	45,900	Shelby (1996)
4	0.0076	L	211	45,900	Shelby (1996)
4	0.0077	SiL	513	54,000	Shelby (1996)
4	0.0079		465	20,600	Shelby (1996)
4	0.0079		1,320	58,400	Shelby (1996)
4	0.0082		2,050	90,700	Shelby (1996)
4	0.0123	SL	384	48,000	Shelby (1996)
4	0.0137	L	71.7	15,600	Shelby (1996)
4	0.0138	SL	363	45,400	Shelby (1996)
4	0.0138	SL	726	90,800	Shelby (1996)
4	0.014	L	57.9	12,600	Shelby (1996)
4	0.0142	L	55.3	12,000	Shelby (1996)
4	0.0148	SiL	251	26,400	Shelby (1996)
4	0.0149	L	210	45,700	Shelby (1996)
4	0.0149	CL	222	19,500	Shelby (1996)
4	0.015	CL	313	27,500	Shelby (1996)
4	0.0151	L	207	45,000	Shelby (1996)
4	0.0152	L	220	47,800	Shelby (1996)
4	0.0154	CL	270	23,700	Shelby (1996)
4	0.0154		616	27,300	Shelby (1996)
4	0.0158	SiL	494	52,000	Shelby (1996)
4	0.016	SiL	364	38,300	Shelby (1996)
4	0.016		2,670	118,000	Shelby (1996)
4	0.0168		800	35,400	Shelby (1996)
4	0.0212	SL	259	32,400	Shelby (1996)
4	0.0217	SL	472	59,000	Shelby (1996)
4	0.022	SL	579	72,400	Shelby (1996)
4	0.0316	L	46.7	10,200	Shelby (1996)

Table 89. (Continued).

Expt rating	Soil conc, ppm	Soil texture	K_d	K_{oc}	Reference
4	0.0342	L	68.8	15,000	Shelby (1996)
4	0.035	L	79.7	17,400	Shelby (1996)
4	0.0353	SiL	692	72,800	Shelby (1996)
4	0.0356	CL	252	22,100	Shelby (1996)
4	0.0366	SiL	832	87,600	Shelby (1996)
4	0.0372	L	210	45,700	Shelby (1996)
4	0.0377		857	37,900	Shelby (1996)
4	0.0379	L	209	45,400	Shelby (1996)
4	0.0381	CL	278	24,400	Shelby (1996)
4	0.0384	SiL	600	63,200	Shelby (1996)
4	0.0384		674	29,800	Shelby (1996)
4	0.0385	SL	353	44,100	Shelby (1996)
4	0.0387	CL	239	21,000	Shelby (1996)
4	0.0396	SL	192	24,000	Shelby (1996)
4	0.0396		861	38,100	Shelby (1996)
4	0.042	SL	294	36,800	Shelby (1996)
4	0.0697	L	207	45,000	Shelby (1996)
4	0.07	L	219	47,600	Shelby (1996)
4	0.0711	L	212	46,100	Shelby (1996)
4	9	L	554	95,200	Daly and Williams (1990)
All ratings	Avg, <0.5		453	41,700	
	CV, %		114.5	75.2	
	Avg, 0.5 to <1				
	CV, %				
	Avg, 1–5				
	CV, %				
	Avg, >5		554	95,200	
	CV, %				
	Avg, All Conc		454[a]	42,500[a]	
	CV, %		113.3	74.8	
>4	Avg, <0.5				
	CV, %				
	Avg, 0.5 to <1				
	CV, %				
	Avg, 1–5				
	CV, %				
	Avg, >5				
	CV, %				
	Avg, All Conc				
	CV, %				

[a]Representative value.

Table 90. Fenpropathrin soil desorption partition coefficients.

Expt rating	Soil conc, ppm	Soil texture	K_{dd}	K_{ocd}	Reference
4	0.0022	CL	275	24,100	Shelby (1996)
4	0.0022	CL	1,100	96,500	Shelby (1996)
4	0.0025	L	69.4	15,100	Shelby (1996)
4	0.0025	L	52.1	11,300	Shelby (1996)
4	0.0026	L	37.7	8,200	Shelby (1996)
4	0.0032	CL	400	35,100	Shelby (1996)
4	0.0033		413	18,300	Shelby (1996)
4	0.0034	SiL	567	59,700	Shelby (1996)
4	0.0034		567	25,100	Shelby (1996)
4	0.0035	SiL	140	14,700	Shelby (1996)
4	0.0041		2,050	90,700	Shelby (1996)
4	0.0049	L	27.7	6,020	Shelby (1996)
4	0.005	L	36	7,830	Shelby (1996)
4	0.0051	L	32.3	7,020	Shelby (1996)
4	0.0064	CL	200	17,500	Shelby (1996)
4	0.0066	L	388	84,300	Shelby (1996)
4	0.0066	SiL	80.5	8,470	Shelby (1996)
4	0.0069	L	363	78,900	Shelby (1996)
4	0.0069	SiL	256	26,900	Shelby (1996)
4	0.007	L	467	100,000	Shelby (1996)
4	0.0072	SiL	343	36,100	Shelby (1996)
4	0.0079	SL	376	47,000	Shelby (1996)
4	0.008	SL	190	23,800	Shelby (1996)
4	0.0083	CL	638	56,000	Shelby (1996)
4	0.0089	SL	424	53,000	Shelby (1996)
4	0.0094	L	37.3	8,110	Shelby (1996)
4	0.0108	L	28.3	6,150	Shelby (1996)
4	0.0113	L	36.8	8,000	Shelby (1996)
4	0.0134	L	372	80,900	Shelby (1996)
4	0.0135	SiL	422	44,400	Shelby (1996)
4	0.0136	L	378	82,200	Shelby (1996)
4	0.0137	CL	180	15,800	Shelby (1996)
4	0.0139	L	409	88,900	Shelby (1996)
4	0.014	CL	292	25,600	Shelby (1996)
4	0.0141		671	29,700	Shelby (1996)
4	0.0142	CL	309	27,100	Shelby (1996)
4	0.0147	SiL	320	33,700	Shelby (1996)
4	0.0152	SiL	211	22,200	Shelby (1996)
4	0.0153		900	39,800	Shelby (1996)
4	0.0166		664	29,400	Shelby (1996)
4	0.017	SL	354	44,300	Shelby (1996)
4	0.0179	SL	303	37,900	Shelby (1996)
4	0.0188	SL	448	56,000	Shelby (1996)

Table 90. (Continued).

Expt rating	Soil conc, ppm	Soil texture	K_{dd}	K_{ocd}	Reference
4	0.0204	L	21.8	4,740	Shelby (1996)
4	0.0208	L	26.3	5,720	Shelby (1996)
4	0.026	CL	310	27,200	Shelby (1996)
4	0.0269	L	31.1	6,760	Shelby (1996)
4	0.0323	CL	278	24,400	Shelby (1996)
4	0.0326		776	34,300	Shelby (1996)
4	0.0334	CL	367	32,200	Shelby (1996)
4	0.0338	L	380	82,600	Shelby (1996)
4	0.0339		997	44,100	Shelby (1996)
4	0.0346	SiL	298	31,400	Shelby (1996)
4	0.0346	SL	336	42,000	Shelby (1996)
4	0.0353	SiL	310	32,600	Shelby (1996)
4	0.0354		1,220	54,000	Shelby (1996)
4	0.0359	L	403	87,600	Shelby (1996)
4	0.0361	SL	410	51,300	Shelby (1996)
4	0.0363	SiL	352	37,100	Shelby (1996)
4	0.0373	SL	296	37,000	Shelby (1996)
4	0.0622	L	360	78,300	Shelby (1996)
4	0.0632	L	365	79,300	Shelby (1996)
4	0.0639	L	453	98,500	Shelby (1996)
All ratings	Avg, <0.5		378	40,000	
	CV, %		88.5	70.1	
	Avg, 0.5 to <1				
	CV, %				
	Avg, 1–5				
	CV, %				
	Avg, >5				
	CV, %				
	Avg, All Conc		378	40,000	
	CV, %		88.5	70.1	
>4	Avg, <0.5				
	CV, %				
	Avg, 0.5 to <1				
	CV, %				
	Avg, 1–5				
	CV, %				
	Avg, >5				
	CV, %				
	Avg, All Conc				
	CV, %				

Table 91. Freundlich sorption data for fenpropathrin.

Experiment rating	Soil texture	K_{df}	K_{ocf}	$1/n$	Reference
4	L	42.5	9,240	0.953	Shelby (1996)
4	SiL	16.8	1,770	0.66	Shelby (1996)
4	SL	20.4	2,560	0.708	Shelby (1996)
4	CL	19	1,670	0.724	Shelby (1996)
4		12.8	567	0.606	Shelby (1996)
4	L	219	47,700	1	Shelby (1996)
All ratings	Avg	55.1	10,600	0.775	
	CV, %	147.0	174.3	20.9	
>4	Avg				
	CV, %				

and 91. Shelby also reran sorption and desorption on one of the soils used originally by Lee, using the same methodology but shortening the equilibration period to 2.5 hr to minimize fenpropathrin breakdown during equilibration. Identity of ^{14}C was not determined in water or soil, and any breakdown of fenpropathrin to more polar material would have contributed to underestimation of K_d or K_{dd}.

Although not a soil sorption study, work by Daly and Williams (1990) from an anaerobic soil degradation experiment provided a single-point measure of fenpropathrin partitioning between soil and water. Measurement of fenpropathrin concentrations in soil and water after 15 d of anaerobic incubation yielded the value cited in Table 89, which is the mean of two replicate samples. All values in Table 89 are associated with uncertainty, and thus the mean K_d and K_{oc} values over all concentrations is selected as the representative value for fenpropathrin. Desorption and Freundlich K_d and K_{oc} values from the report by Shelby (1996) are presented in Tables 90 and 91, respectively.

B. Abiotic Chemical Properties

1. Hydrolysis.   Hydrolysis of fenpropathrin in sterile buffered water was characterized by Takahashi et al. in 1983 (1983b) and again by Concha et al. in 1992 (1992b). Table 92 summarizes their results as nonlinear and linear first-order rate constants at each pH. The Takahashi reference cited in the table contained no raw data, only linear first-order plots, so nonlinear analysis was not possible. With regard to property representation, the Concha data are selected over the Takahashi data because of the latter's absence of raw data.

2. Photolysis in Water.   Two reports were reviewed for photolysis of fenpropathrin in water (Table 93). An early study by Takahashi et al. (1983a) utilized

Table 92. Hydrolysis of fenpropathrin in buffered water.

Temp, °C	pH	Nonlinear		Linear		Reference
		k^a	Half-life, d^a	k^a	Half-life, d^a	
25	5	$4.15E-04^b$	1,670	$4.83E-04$	1,440	Concha et al. (1992b)
25	7	$1.25E-03^b$	555	$1.22E-03$	568	Concha et al. (1992b)
25	7			$6.13E-04$	1,130	Takahashi et al. (1983b)
	Avg CV, %			$9.17E-04$ 46.8		
25	9	$4.83E-02^b$	14.4	$4.47E-02$	15.5	Concha et al. (1992b)
25	9			$5.02E-02$	13.8	Takahashi et al. (1983b)
	Avg CV, %			$4.74E-02$ 8.2		

[a]An entry of $0.00E+00$ means k was measured but fenpropathrin is stable; a blank k or half-life value means no measurement.
[b]Representative value.

Tween 85 as an emulsifier to prepare solutions of fenpropathrin that were then exposed to natural sunlight. The more recent study by Jalai-Araghi et al. (1992) used buffer plus a small amount of acetonitrile to prepare solutions that were then exposed to California natural sunlight up to 30 d. Dark controls were included and used to correct the kinetics data reported in Table 93 for solution degradation that occurred in the absence of light. Data from the Jalai-Araghi report are selected to represent fenpropathrin because of the presence of emulsifier in the Takahashi study.

Table 93. Photolysis of fenpropathrin in water.

Natural sun	Buffer medium	Nonlinear		Linear		Reference
		k	Half-life, d	k	Half-life, d	
Yes	No	$8.40E-03$	82.5	$7.92E-03$	87.5	Takahashi et al. (1983a)
Yes	No	$4.19E-02$	16.5	$4.75E-02$	14.6	Takahashi et al. (1983a)
Yes	Yes	$1.15E-03^a$	603	$1.08E-03$	642	Jalai-Araghi et al. (1992)

[a]Representative value.

3. Photolysis on Soil. Takahashi et al. (1983a) applied fenpropathrin in diethyl-ether to the surfaces (at 15 ppm) of three soils coated on glass plates, and then exposed the plates in an air-dry state to Japanese natural sunlight up to 14 d (Table 94). Concha et al. (1992a) applied chemical at 120 ppm to soil surfaces kept wet during irradiation by California sunlight for periods up to 30 d. Both studies carried dark controls so that data in Table 94 are corrected for dark reaction. The study carried out by Takahashi et al. at lower concentration and with air-dry soil produced rapid fenpropathrin photolysis; the study by Concha et al. at higher concentration and wet soil indicated little or no breakdown. Since soil surfaces are most often air dry but occasionally can be wet when the sun shines, it was decided to combine all data through a common average and allow the average to represent the photolysis of fenpropathrin on soil.

C. Biotic Chemical Properties

1. Aerobic Soil Degradation. Aerobic soil degradation studies have been con-ducted by Mikami et al. (1983, 1984b), Roberts and Standen (1976), Cranor (1990), and Daly and Williams (1990) at different concentrations in a variety of soils. Table 95 summarizes results from nonlinear and linear kinetic analysis of raw data from each report.

The first study by Roberts and Standen (1976) was carried out at fairly high application rate and in the presence of a somewhat high volume of solvent (250 μL acetone/50 g soil). The second and third studies by Mikami et al. (1983, 1984b) used less solvent (100 μL acetone/30 g soil) and lower application rates. The fourth study, by Cranor (1990), was conducted at a high application rate of 10 ppm of fenpropathrin added in 25 mL acetone to 357.5 g soil (3.5 mL/50 g soil, allowed to evaporate during mixing). The fifth study also utilized a high

Table 94. Photolysis of fenpropathrin on soil.

Natural sun	Moisture state	Nonlinear		Linear		Reference
		k	Half-life, d	k	Half-life, d	
Yes	Air dry	4.26E − 01	1.63	1.84E − 01	3.77	Takahashi et al. (1983a)
Yes	Air dry	1.02E − 01	6.8	7.57E − 02	9.16	Takahashi et al. (1983a)
Yes	Air dry	9.33E − 02	7.43	7.36E − 02	9.42	Takahashi et al. (1983a)
Yes	75% 1/3 bar	0.00E + 00	0	0.00E + 00	0	Concha et al. (1992a)
	Avg	*1.55E − 01*[a]				
	CV, %	119.9				

[a]Representative value.

Table 95. Fenpropathrin degradation in aerobic soil.

Expt rating	Temp, °C	Soil conc, ppm	Soil texture	Nonlinear k	Nonlinear Half-life, d	Linear k	Linear Half-life, d	Reference
6	25	1	SCL	2.17E − 02	31.9	2.03E − 02	34.1	Mikami et al. (1984b)
6	25	1	C	2.21E − 02	31.4	2.16E − 02	32.1	Mikami et al. (1984b)
7	25	1	SCL	7.38E − 02	9.39	1.99E − 02	34.8	Mikami et al. (1983)
7	25	1	C	4.15E − 02	16.7	1.90E − 02	36.5	Mikami et al. (1983)
3	24	2.86	SL	1.91E − 02	36.3	1.76E − 02	39.4	Roberts and Standen (1976)
3	24	2.86	C	5.46E − 03	127	5.22E − 03	133	Roberts and Standen (1976)
3	24	2.86	SL	2.65E − 02	26.2	2.00E − 02	34.7	Roberts and Standen (1976)
3	24	2.86	SC	1.71E − 02	40.5	1.35E − 02	51.3	Roberts and Standen (1976)
3	24	2.86	SC	3.87E − 03	179	3.14E − 03	221	Roberts and Standen (1976)
2	25	10	SiL	4.47E − 03	155	4.55E − 03	152	Cranor (1990)
2	25	12	L	3.74E − 03	185	3.73E − 03	186	Daly and Williams (1990)
All ratings		Avg, <0.5						
		CV, %						
		Avg + CL						
		Avg, 0.5–1.5			22.3		34.4	
		CV, %			49.9		5.3	
		Avg + CL			31.5		35.9	
		Avg, 1.15–10			94.0		105	
		CV, %			71.9		71.8	
		Avg + CL			135		151	
		Avg, >10			185		186	
		CV, %						
		Avg + CL			555		558	
		Avg, All Conc			76.2		86.8	
		CV, %			91.4		82.6	
		Avg + CL			105		116	
>4		Avg, All Conc			22.3[a]		34.4	
		CV, %			49.9		5.3	
		Avg + CL			31.5[a]		35.9	

[a]Representative value.

application rate and large acetone volume (4.1 mL/50 g soil). The use of large volumes of solvent and high application rates in three of the five studies casts some uncertainty on the kinetic results from these studies, and the experiments were rated at a lower level of confidence.

2. Anaerobic Soil Degradation. Mikami et al. (1983) added fenpropathrin in 100 μL acetone to 30 g soil in beakers incubated in dessicators containing pyrogallol to sorb oxygen and purged continuously with nitrogen gas. Samples were maintained at 40% WHC (water-holding capacity) and thus were not saturated or protected from oxygen by a barrier of water over the soil surface.

Daly and Williams examined anaerobic degradation again in 1990, with a similar procedure used for their aerobic studies described earlier. High levels of acetone were used (4.1 mL/50 g soil) to apply a high application rate (12 ppm) to samples incubated aerobically at 75% 1/3 bar moisture for the first 30 d and then anaerobically by waterlogging, adding glucose to stimulate soil reduction, and by changing the incubation atmosphere to nitrogen. Results from these experiments are summarized in Table 96.

3. Aerobic Aquatic Degradation. No aerobic aquatic information was reviewed for fenpropathrin.

Table 96. Fenpropathrin degradation generated by an anaerobic soil experimental design.

Expt rating	Temp, °C	Soil conc, ppm	Soil texture	Nonlinear k	Nonlinear Half-life, d	Linear k	Linear Half-life, d	Reference
4	25	1	SCL	1.84E − 03	377	1.81E − 03	383	Mikami et al. (1983)
4	25	1	C	2.51E − 03	276	2.45E − 03	283	Mikami et al. (1983)
3	25	12	L	3.95E − 03	175	3.78E − 03	183	Daly and Williams (1990)
All ratings	Avg, <1.5				327		333	
	CV, %				21.9		21.2	
	Avg + CL				482		487	
	Avg, >1.5				175		183	
	CV, %							
	Avg + CL				525		549	
	Avg, All Conc				276[a]		283	
	CV, %				36.6		35.3	
	Avg + CL				386[a]		392	
>4	Avg, All Conc							
	CV, %							
	Avg + CL							

[a]Representative value.

4. Anaerobic Aquatic Degradation. Roberts and Standen (1976) added fenpropathrin in 250 μL acetone/50 g soil, waterlogged the soil to provide the water barrier, and incubated the samples under nitrogen gas for alternating periods of light and dark for up to 160 d. Results are summarized in Table 97, but it is noted that the use of light in the experiment lowers the level of confidence in the results from this study. Kaman (1994, 1995) reexamined anaerobic degradation by adding fenpropathrin in 95 μL acetonitrile to 10 g soil + 20 mL water incubated in dessicators kept dark and purged continuously with oxygen-free nitrogen gas. These results are also summarized in Table 97. Data used to represent fenpropathrin were selected from Kaman's work because of its higher experimental rating.

IX. Lambda-cyhalothrin

Cyhalothrin is [1α,3α(Z)-(±)-cyano-(3-phenoxyphenyl)methyl 3-(2-chloro-3,3,3-trifluoro-1-propenyl)-2,2-dimethylcyclopropanecarboxylate by the C.A. naming system and (RS)-α-cyano-3-phenoxybenzyl (Z)-(1RS,3RS)-(2-chloro-3,3,3-trifluoropropenyl)-2,2-dimethylcyclopropanecarboxylate by the IUPAC system. It consists of a mixture of four stereoisomers, all with the *cis* configuration about the

Table 97. Fenpropathrin degradation generated by an anaerobic aquatic experimental design.

Expt rating	Temp, °C	Soil conc, ppm	Soil texture	Nonlinear k	Nonlinear Half-life, d	Linear k	Linear Half-life, d	Reference
7	25	0.4	L	1.02E − 02	68	7.42E − 03	93.4	Kaman (1994)
7	25	0.4	L	8.75E − 03	79.2	8.15E − 03	85	Kaman (1995)
4	25	2.5	SL	1.43E − 02	48.5	1.20E − 02	57.8	Roberts and Standen (1976)
All ratings	Avg, <1.5				73.6		89.2	
	CV, %				10.8		6.7	
	Avg + CL				90.8		102	
	Avg, >1.5				48.5		57.8	
	CV, %							
	Avg + CL				146		173	
	Avg, All Conc				65.2		78.7	
	CV, %				23.8		23.6	
	Avg + CL				82.2		99.0	
>4	Avg, All Conc				*73.6*[a]		89.2	
	CV, %				10.8		6.7	
	Avg + CL				*90.8*[a]		102	

[a]Representative value.

1,3-position on the cyclopropane ring and Z configuration about the differently substituted terminal olefinic bond. The four stereoisomers comprise the following two diastereoisomeric pairs of enantiomers in the ratio of approximately 60: 40: (1R,3R,alphaR) and (1S,3S,alphaS) known as pair A, and (1R, 3R,alphaS) and (1S,3S,alphaR) known as pair B.

Lambda-cyhalothrin is a 1:1 mixture of the enantiomers of pair B and has a C.A. name of [1α(S*),3α(Z)]-(±)-cyano(3-phenoxyphenyl)methyl 3-(2-chloro-3,3,3-trifluoro-1-propenyl)-2,2-dimethylcyclopropanecarboxylate. The IUPAC names of the pair B enantiomers are (S)-α-cyano-3-phenoxybenzyl (Z)-(1R,3R)-3-(2-chloro-3,3,3-trifluoropropenyl)-2,2-dimethylcyclopropanecarboxylate and (R)-α-cyano-3-phenoxybenzyl (Z)-(1S,3S)-3-(2-chloro-3,3,3-trifluoropropenyl)-2,2-dimethylcyclopropanecarboxylate. Some data on aqueous photolysis and soil degradation of cyhalothrin and the enantiomer pair A have been included in this review for lambda-cyhalothrin. Lambda-cyhalothrin has a CAS RN of 91465–08-6 and has been referred to by PP321, ICIA0321, and OMS 3021 (Tomlin 1994). Cyhalothrin has been referred to as PP563, ICI 146 814, OMS 2011, and has a CAS RN of 68085-85-8.

A. Physical Properties

1. Vapor Pressure.   Vapor pressure of lambda-cyhalothrin (Table 98) has been measured by Wollerton and Husband (1988a) with a gas saturation method using a gas saturation generator column. N_2 gas was passed through the column held at 60°, 70°, or 80 °C and then into a cold trap, which was analyzed gas chromatographically for parent material. Vapor pressure was extrapolated to 20 °C from regression analysis of the results obtained at the higher temperatures and it is this value that is cited in Table 98.

2. Water Solubility.   Water solubility was measured at 20 °C by Wollerton and Husband (1988a) by a generator column method that passes water through a column filled with Chromasorb packing coated with lambda-cyhalothrin. The chemical in column effluent was collected by passage through a u Bondapak C18 column, which was subjected to reverse-phase HPLC analysis. This methodology minimized the chances for formation of stable suspensions during saturation and provides the best opportunity for accurate measurement of pyrethroid water solubility. A total of 12 determinations were made, and the mean of these measurements is reported in Table 99.

Table 98. Vapor pressure of lambda-cyhalothrin.

Temperature, °C	Best values	VP, mm Hg	Source
20		*1.56E – 09*[a]	Wollerton and Husband (1988a)

[a]Representative value.

Table 99. Water solubility of lambda-cyhalothrin.

Temperature, °C	Column saturation	WS, ppm	Reference
20	Yes	*5.00E – 03*[a]	Wollerton and Husband (1988a)

[a]Representative value.

3. Henry's Law Constant.   Henry's law constant, K_h, for lambda-cyhalothrin is calculated from Eq. 1, using a molecular weight of 449.9, a vapor pressure of 1.56×10^{-9} from Table 98, and a water solubility of 5.00×10^{-3} from Table 99. K_h for lambda-cyhalothrin equals 1.9×10^{-7} atm m^3 mol^{-1}.

4. Octanol–Water Partition Coefficient.   Wollerton and Husband (1988a) measured the octanol–water partition coefficient for lambda-cyhalothrin by means of a generator column technique whose principle is similar to that used for vapor pressure and water solubility. Chromasorb column packing coated with lambda-cyhalothrin in octanol was eluted with octanol-saturated water, minimizing the chances for emulsion formation during equilibration or contamination of water phase with trace quantities of octanol. The mean value from eight measurements is presented in Table 100. Log P calculated from the molecular structure of lambda-cyhalothrin is also presented in the table, and its value is slightly less than the measured value.

5. Bioconcentration Factor.   A study by Yamauchi (1985) was conducted with carp exposed continuously to lambda-cyhalothrin in a flow-through system up to 28 d. Periodically water was analyzed for lambda-cyhalothrin levels and fish for total ^{14}C content. BCF was calculated from the ratio of the two. Table 101

Table 100. Octanol–water partition coefficient for lambda-cyhalothrin.

Best values	K_{ow}	Log P	CLog P	Reference
Yes	*1.00E + 07*[a]	7.00[a]	6.1[b]	Wollerton and Husband (1988a)

[a]Representative value.
[b]Calculated from molecular structure by Briggs (1999).

Table 101. Lambda-cyhalothrin bioconcentration factor based on fish whole-body analysis.

Type of experiment	Best values	BCF	Reference
Flow-through	Yes	2240[a]	Yamauchi (1985)

[a]Representative value.

presents the BCF calculated from the lambda-cyhalothrin concentrations measured in the water and total ^{14}C burden in whole-body fish. The value in the table represents the maximum BCF from a range of 1660 to 2240 observed in the study.

6. *Soil Sorption.*   Sorption and desorption of lambda-cyhalothrin in a variety of soils and at different concentrations was carried out by Muller et al. (1996). Various concentrations of ^{14}C-labeled lambda-cyhalothrin in dilute calcium chloride solutions were equilibrated with soils, and then measurements were made of total ^{14}C remaining in the water phase after equilibration. Soil-phase concentration was calculated from the difference in water-phase ^{14}C concentration before and after addition of soil, assuming the difference is due to sorption to soil after correction of sorption to container walls. Material other than lambda-cyhalothrin in the water phase, such as breakdown products, would cause underestimation of sorption.

Desorption was carried out by resuspension of samples in clean calcium chloride after the adsorption phase had been completed, and then by following a procedure similar to that previously described. Adsorption partition coefficients from this study are summarized in Table 102, desorption coefficients in Table 103, and Freundlich sorption coefficients in Table 104.

B. Abiotic Chemical Properties

1. *Hydrolysis.*   Collis and Leahey (1984) measured hydrolysis rates of ^{14}C-labeled lambda-cyhalothrin in sterile, buffered water incubated at 25 °C in the dark. Kinetic data from nonlinear and linear first-order regression analysis of the experimental results are presented in Table 105.

2. *Photolysis in Water.*   Hall and Leahey (1983) studied the fate of cyhalothrin in test systems containing natural river water and sediment incubated in quartz vessels and exposed to natural sunlight in the U.K. (Table 106). Cyhalothrin was applied as a 10% emulsifiable concentrate at a very high initial nominal rate (9.00E – 01 ppm) and far in excess of the compound aqueous solubility. Dark controls were included in the study. Although sunlight is the source of irradiation, the presence of sediment and use of emulsifier provide a level of uncertainty to the results of these studies.

A second study by Curl et al. (1984a) under natural sunlight outdoors used sterile buffer containing a small amount of acetonitrile as cosolvent. Sterile dark controls were included so that kinetics could be corrected for breakdown other than that resulting from photolytic reactions. Samples were analyzed only at 30 d, providing a single sampling point for calculation of first-order kinetics, and this paucity of sampling points causes some uncertainty in the kinetic results of the study.

Photolysis was studied a third time by Priestley and Leahey in 1988 in the laboratory under xenon light filtered to match simulated universal natural light

Table 102. Lambda-cyhalothrin soil adsorption coefficients.

Expt rating	Soil conc, ppm	Soil texture	K_d	K_{oc}	Reference
6	1	LS	2,210	224,000	Muller et al. (1996)
6	1	LS	3,810	386,000	Muller et al. (1996)
6	1	S	2,120	457,000	Muller et al. (1996)
6	1	SL	3,700	83,800	Muller et al. (1996)
6	1	SL	6,070	243,000	Muller et al. (1996)
6	1	SiCL	4,960	231,000	Muller et al. (1996)
6	1	SiCL	5,930	301,000	Muller et al. (1996)
6	1	SL	3,380	307,000	Muller et al. (1996)
6	1	LS	1,870	640,000	Muller et al. (1996)
6	1	SL	5,860	230,000	Muller et al. (1996)
6	2	LS	3,240	328,000	Muller et al. (1996)
6	2	S	2,870	619,000	Muller et al. (1996)
6	2	SL	9,190	360,000	Muller et al. (1996)
6	2	LS	2,130	734,000	Muller et al. (1996)
6	2	LS	1,620	164,000	Muller et al. (1996)
6	2	SL	3,710	336,000	Muller et al. (1996)
6	2	SiCL	5,190	242,000	Muller et al. (1996)
6	2	SL	6,250	142,000	Muller et al. (1996)
6	2	SL	5,120	205,000	Muller et al. (1996)
6	2	SiCL	5,700	289,000	Muller et al. (1996)
6	5	LS	1,640	166,000	Muller et al. (1996)
6	5	LS	3,860	391,000	Muller et al. (1996)
6	5	SiCL	4,280	199,000	Muller et al. (1996)
6	5	LS	2,310	797,000	Muller et al. (1996)
6	5	SL	8,930	358,000	Muller et al. (1996)
6	5	S	1,830	394,000	Muller et al. (1996)
6	5	SL	8,290	325,000	Muller et al. (1996)
6	5	SL	5,940	539,000	Muller et al. (1996)
6	5	SiCL	6,060	307,000	Muller et al. (1996)
6	5	SL	7,120	161,000	Muller et al. (1996)
6	9	LS	2,520	255,000	Muller et al. (1996)
6	9	SiCL	6,150	312,000	Muller et al. (1996)
6	9	LS	2,300	234,000	Muller et al. (1996)
6	9	S	2,580	555,000	Muller et al. (1996)
6	9	SiCL	3,380	158,000	Muller et al. (1996)
6	9	LS	2,420	836,000	Muller et al. (1996)
6	10	SL	3,080	69,900	Muller et al. (1996)
6	10	SL	2,830	257,000	Muller et al. (1996)
6	10	SL	7,700	302,000	Muller et al. (1996)
6	10	SL	7,330	294,000	Muller et al. (1996)
6	18	LS	2,070	210,000	Muller et al. (1996)
6	18	LS	1,770	609,000	Muller et al. (1996)
6	18	LS	3,920	398,000	Muller et al. (1996)
6	18	S	2,630	566,000	Muller et al. (1996)

Table 102. (Continued.)

Expt rating	Soil conc, ppm	Soil texture	K_d	K_{oc}	Reference
6	19	SiCL	4,640	216,000	Muller et al. (1996)
6	19	SL	4,190	95,100	Muller et al. (1996)
6	19	SL	3,390	133,000	Muller et al. (1996)
6	19	SiCL	5,550	282,000	Muller et al. (1996)
6	19	SL	10,600	425,000	Muller et al. (1996)
6	20	SL	3,180	289,000	Muller et al. (1996)
All	Avg, <0.5				
ratings	CV, %				
	Avg, 0.5 to <1				
	CV, %				
	Avg, 1–5		4,250	326,000	
	CV, %		45.7	52.2	
	Avg, >5		4,420	338,000	
	CV, %		54.4	56.1	
	Avg, All Conc		4,350	333,000	
	CV, %		50.9	54.1	
>4	Avg, <0.5				
	CV, %				
	Avg, 0.5 to <1				
	CV, %				
	Avg 1–5		4,250[a]	326,000[a]	
	CV, %		45.7	52.2	
	Avg, >5		4,420	338,000	
	CV, %		54.4	56.1	
	All Conc		4,350	333,000	
	CV, %		50.9	54.1	

[a]Representative value.

of D65 radiation. Light intensity at an equivalent distance to the sample surface was monitored, and results were converted to equivalent radiation days of autumn sun in Florida, assuming 12 hr of sunlight per day. Sterile dark controls showed no degradation or loss of lambda-cyhalothrin during the course of the study. A number of replicated sampling periods provided quality data for kinetic analysis.

A fourth study by Moffatt (1994b) measured quantum yield of lambda-cyhalothrin using xenon light from an apparatus designed specifically to measure quantum yield at 285 nm. Quantum yield was calculated as 9.2×10^{-2}, which was then used to estimate the photolysis rate constant and half-life cited in Table 106, assuming mid-European spring sun intensity and clean water depth of 5 cm.

Table 103. Lambda-cyhalothrin soil desorption partition coefficients.

Expt rating	Soil conc, ppm	Soil texture	K_{dd}	K_{ocd}	Reference
6	1	LS	1,380	140,000	Muller et al. (1996)
6	1	LS	2,950	299,000	Muller et al. (1996)
6	1	S	2,880	620,000	Muller et al. (1996)
6	1	SL	5,320	121,000	Muller et al. (1996)
6	1	SL	10,600	425,000	Muller et al. (1996)
6	1	SiCL	4,870	247,000	Muller et al. (1996)
6	1	SiCL	4,000	186,000	Muller et al. (1996)
6	1	LS	1,720	593,000	Muller et al. (1996)
6	1	SL	4,670	423,000	Muller et al. (1996)
6	1	SL	9,950	390,000	Muller et al. (1996)
6	2	LS	3,690	374,000	Muller et al. (1996)
6	2	SL	9,440	370,000	Muller et al. (1996)
6	2	LS	1,750	602,000	Muller et al. (1996)
6	2	LS	1,460	148,000	Muller et al. (1996)
6	2	S	2,270	490,000	Muller et al. (1996)
6	2	SL	3,760	341,000	Muller et al. (1996)
6	2	SL	4,360	98,800	Muller et al. (1996)
6	2	SL	10,400	417,000	Muller et al. (1996)
6	2	SiCL	4,990	253,000	Muller et al. (1996)
6	2	SiCL	4,860	226,000	Muller et al. (1996)
6	5	SiCL	3,950	184,000	Muller et al. (1996)
6	5	LS	6,750	684,000	Muller et al. (1996)
6	5	SL	6,770	271,000	Muller et al. (1996)
6	5	LS	3,190	1,100,000	Muller et al. (1996)
6	5	S	3,200	690,000	Muller et al. (1996)
6	5	LS	1,790	181,000	Muller et al. (1996)
6	5	SL	10,290	403,000	Muller et al. (1996)
6	5	SL	3,830	348,000	Muller et al. (1996)
6	5	SiCL	5,000	253,000	Muller et al. (1996)
6	5	SL	3,370	76,400	Muller et al. (1996)
6	9	LS	3,460	1,190,000	Muller et al. (1996)
6	9	LS	1,490	151,000	Muller et al. (1996)
6	9	SiCL	4,580	232,000	Muller et al. (1996)
6	9	LS	4,360	442,000	Muller et al. (1996)
6	9	S	2,550	550,000	Muller et al. (1996)
6	10	SL	8,730	350,000	Muller et al. (1996)
6	10	SiCL	5,100	238,000	Muller et al. (1996)
6	10	SL	12,180	477,000	Muller et al. (1996)
6	10	SL	4,190	380,000	Muller et al. (1996)
6	10	SL	5,410	123,000	Muller et al. (1996)
6	18	LS	3,780	383,000	Muller et al. (1996)
6	18	LS	1,960	199,000	Muller et al. (1996)
6	18	LS	3,720	1,280,000	Muller et al. (1996)
6	18	S	2,560	551,000	Muller et al. (1996)

Table 103. (Continued).

Expt rating	Soil conc, ppm	Soil texture	K_{dd}	K_{ocd}	Reference
6	19	SL	6,680	268,000	Muller et al. (1996)
6	19	SiCL	6,350	322,000	Muller et al. (1996)
6	19	SiCL	4,360	203,000	Muller et al. (1996)
6	20	SL	3,560	80,700	Muller et al. (1996)
6	20	SL	8,790	344,000	Muller et al. (1996)
6	20	SL	6,380	579,000	Muller et al. (1996)
All ratings	Avg, <0.5 ppm				
	CV, %				
	Avg, 0.5–1 ppm				
	CV, %				
	Avg, 1–5 ppm		4,770	338,000	
	CV, %		63.1	47.8	
	Avg, >5 ppm		4,940	418,000	
	CV, %		51.1	74.0	
	Avg, All Conc		4,870	386,000	
	CV, %		55.4	67.7	
>4	Avg, <0.5 ppm				
	CV, %				
	Avg, 0.5–1 ppm				
	CV, %				
	Avg 1–5		4,770	338,000	
	CV, %		63.1	47.8	
	Avg, >5 ppm		4,940	418,000	
	CV, %		51.1	74.0	
	All Conc		4,870	386,000	
	CV, %		55.4	67.7	

3. *Photolysis on Soil.* Two studies of the photolysis of lambda-cyhalothrin on soil surfaces have been carried out (Table 107); the first was by Curl et al. (1984b) and the second by Parker and Leahey (1986). The first study irradiated lambda-cyhalothrin on dry soil plates with light from either a xenon arc lamp in the laboratory or from natural sunlight in an outdoor setting. Dark controls were included, and data presented in Table 107 reflect the dark control correction. The indoor study was sampled several times so that regression analysis could be performed on the data; irradiation was continuous and so corrections were made for light/dark periods of 12 hr each, in addition to corrections making the artificial light match that of sunlight. The sunlight study was sampled only at 30 d, so the first-order point equation (Eq. 4) was used to calculate an apparent rate constant over the 30-d period.

The second study (Parker and Leahey 1986) examined photolysis in the labo-

Table 104. Freundlich sorption data for lambda-cyhalothrin.

Expt rating	Soil texture	K_{df}	K_{ocf}	$1/n$	Reference
6	SiCL	5,440	276,000	0.99	Muller et al. (1996)
6	LS	2,080	211,000	1.01	Muller et al. (1996)
6	LS	2,560	260,000	0.96	Muller et al. (1996)
6	SL	33,000	1,320,000	1.21	Muller et al. (1996)
6	SiCL	2,360	110,000	0.91	Muller et al. (1996)
6	LS	1,960	676,000	0.99	Muller et al. (1996)
6	S	2,520	543,000	1.01	Muller et al. (1996)
6	SL	1,500	588,000	0.8	Muller et al. (1996)
6	SL	1,780	162,000	0.89	Muller et al. (1996)
6	SL	1,660	37,700	0.85	Muller et al. (1996)
All ratings	Avg	5,490	4,180	0.962	
	CV, %	177.4	91.5	11.7	
>4	Avg	5,490	4,180	0.962	
	CV, %	177.4	91.5	11.7	

ratory with light from a xenon lamp, but this time the lamp was turned on and off in a 24-hr period to simulate a day period of 7 hr light followed by 17 hr dark. In this experiment data from the analysis of the dark controls at the end of 35 d indicated that degradation was more extensive in the dark (26%) than in irradiated samples (13%–16%). Although all soil irradiation plates started out air dry, the faster rate of degradation in the controls was attributed to higher moisture level in dark controls than in irradiated samples, leading to chemical hydrolysis of the cyanide group in the control samples. Because one of the

Table 105. Hydrolysis of lambda-cyhalothrin in buffered water.

Temp, °C	pH	Nonlinear		Linear		Reference
		k[a]	Half-life, d	k[a]	Half-life, d	
25	5	*0.00E + 00*[b]	0	0.00E + 00	0	Collis and Leahey (1984)
25	7	*0.00E + 00*[b]	0	0.00E + 00	0	Collis and Leahey (1984)
25	9	*8.00E − 02*[b]	8.66	5.30E − 02	13.1	Collis and Leahey (1984)

[a]An entry of 0.00E + 00 means k was measured but lambda-cyhalothrin is stable.
[b]Representative value.

Table 106. Photolysis of lambda-cyhalothrin in water.

Natural sun	Buffer medium	Nonlinear		Linear		Reference
		k	Half-life, d	k	Half-life, d	
No	Yes	$2.83E - 02$[a]	24.5	$2.81E - 02$	24.7	Priestley and Leahey (1988)
No	Yes			$1.93E - 01$	3.6	Moffatt (1994b)
Yes	Yes			$1.35E - 02$	51.3[b]	Curl et al. (1984a)
Yes	No	$2.18E - 02$	31.8	$2.03E - 02$	34.1	Hall and Leahey (1983)
Yes	No	$3.27E - 02$	21.2	$3.62E - 02$	19.1	Hall and Leahey (1983)

[a]Representative value.
[b]Estimated from first-order point Eq. 3.

artificial light experiments indicates fairly rapid photolysis and the other no photolysis, the single experiment in natural sunlight is chosen to represent the photolysis characteristics of lambda-cyhalothrin on soil.

C. Biotic Chemical Properties

1. Aerobic Soil Degradation. Bharti et al. (1985) added separately ^{14}C-labeled cyhalothrin, lambda-cyhalothrin, and the cyhalothrin enantiomer pair A in acetone to soil at 40% water-holding capacity (WHC) (50–100 μL acetone solution/ 25 g oven-dry soil) providing initial soil concentrations of the test chemicals of 0.46–2.3 ppm. The higher rate exceeds expected soil concentrations resulting from normal use of lambda-cyhalothrin. The breakdown of the parent compounds was followed at 20 °C during a 180-d period. First-order rate constants and half-lives from nonlinear and linear regression analysis of the data are presented in Table 108.

Table 107. Photolysis of lambda-cyhalothrin on soil.

Natural sun	Moisture state	Nonlinear		Linear		Reference
		k	Half-life, d	k	Half-life, d	
No	Air dry	$2.12E - 01$	3.27	$5.00E - 02$	13.9	Curl et al. (1984b)
Yes	Air dry			$1.29E - 02$[a]	53.7	Curl et al. (1984b)
No	Air dry			$0.00E + 00$	0	Parker and Leahey (1986)

[a]Representative value; estimated from first-order point Eq. 3.

Table 108. Rate of lambda-cyhalothrin degradation in aerobic soil.

Expt rating	Temp, °C	Soil conc, ppm	Soil texture	Nonlinear		Linear		Reference
				k	Half-life, d	k	Half-life, d	
7	20	0.46	SL	2.79E − 02	24.8	2.39E − 02	29	Bharti et al. (1985)
7	20	0.46	LS	8.48E − 03	81.7	8.35E − 03	83	Bharti et al. (1985)
7	20	0.46	SL	2.98E − 02	23.3	1.58E − 02	43.9	Bharti et al. (1985)
7	20	0.47	SL	1.89E − 02	36.7	1.50E − 02	46.2	Bharti et al. (1985)
7	20	2.3	SL	1.49E − 02	46.5	1.44E − 02	48.1	Bharti et al. (1985)
All ratings		Avg, <0.5			41.6		50.5	
		CV, %			65.8		45.4	
		Avg + CL			64.0		69.3	
		Avg, 0.5–1.5						
		CV, %						
		Avg + CL						
		Avg, 1.5–10			46.5		48.1	
		CV, %						
		Avg + CL			140		144	
		Avg, 10						
		CV, %						
		Avg + CL						
		Avg, All Conc			42.6		50.0	
		CV, %			55.9		39.8	
		Avg + CL			58.9		63.7	
>4		Avg, All Conc			42.6[a]		50.0	
		CV, %			55.9		39.8	
		Avg + CL			58.9[a]		63.7	

[a]Representative value.

2. Anaerobic Soil Degradation. No studies were reviewed that utilized an anaerobic soil experimental design for lambda-cyhalothrin.

3. Aerobic Aquatic Degradation. Marriott et al. (1998) investigated the degradation of lambda-cyhalothrin in preincubated test systems composed of natural water and sediment from two sources with differing organic matter content and texture. Lambda-cyhalothrin in acetone (100 μL) was applied to the water surface in each system, and systems were then incubated in the dark at 20 °C for up to 98 d. Aeration was a continuous stream of air passed through air flow-through glass vessels. Results from nonlinear and linear regression analysis of the data collected at various sampling intervals are shown in Table 109.

Table 109. Degradation of lambda-cyhalothrin in aerobic aquatic systems.

Sample name	Sediment concentration, ppm	Nonlinear		Linear		Reference
		k	Half-life, d	k	Half-life, d	
Old Basing	0.02	3.25E − 02	21.3	2.03E − 02	34.1	Marriott et al. (1998)
Virginia Water	0.012	5.49E − 02	12.6	3.29E − 02	21.1	Marriott et al. (1998)
Old Basing	0.2	2.21E − 02	31.4	2.00E − 02	34.7	Marriott et al. (1998)
Virginia Water	0.12	3.10E − 02	22.4	3.30E − 02	21	Marriott et al. (1998)
Avg			21.9[a]		27.7	
CV, %			35.1		27.8	
Avg + CL			28.2[a]		34.0	

[a]Representative value.

4. Anaerobic Aquatic Degradation.   No information was available for the degradation of lambda-cyhalothrin anaerobically in experiments that utilized an anaerobic aquatic experimental design.

X. Permethrin

Permethrin has a C.A. name of (3-phenoxyphenyl)methyl 3-(2,2-dichloro-
ethenyl)-2,2-dimethylcyclopropanecarboxylate (IUPAC = 3-phenoxybenzyl (1*RS*,
3*RS*;1*RS*,3*SR*)-3-(2,2-dichlorovinyl)-2,2-dimethylcyclopropanecarboxylate) and
a molecular weight of 391.3. Typically it is a mixture of cis and trans isomers
in the ratio of 52/48. It has been referred to as NRDC 143, FMC 33297, PP557,
WL 43479, and LE 79-519. CAS RN is 52645-53-1; Official Code is OMS
1821.

A. Physical Properties

1. Vapor Pressure.   Vapor pressures shown in Table 110 have been measured
by the gas saturation method at temperatures between 40° and 60 °C, with extrapolation to 25 °C by the Clausius–Clapeyron equation (Alvarez 1989). The
first value in Table 110 is the vapor pressure of the cis isomer; the second value
is for trans; and the third is the vapor pressure for permethrin according to
Raoult's law, assuming permethrin is a homogeneous liquid comprised of 52%
cis and 48% *trans* isomers and purity greater than 95%. Permethrin vapor pressure is the sum of the two partial pressures according to the following calculation:

$$Vp = (0.52 \times 2.15E - 8) + (0.48 \times 6.90E - 9) = 1.48 \times 10^{-8} \text{ mm Hg}$$

2. Water Solubility.   Permethrin solubility in water has been measured by Wollerton (1987) and by Alvarez (1989). The work by Wollerton is referenced as a
brief summary that indicates a generator column technique was used to provide
saturated solutions, but no details or supporting data are provided other than
values at pH 5, 7, and 9. These solubilities are listed in Table 111, along with
their average and coefficient of variation.

   Alvarez (1989) coated the walls of vials with *cis*- or *trans*-permethrin, added
water and a magnetic stirring bar, and then heated the vials to 50 °C with stir-

Table 110. Permethrin vapor pressure.

Temp, °C	Best value	VP, mm Hg	Source
25	Yes	2.15E − 08	Alvarez (1989)
25	Yes	6.90E − 09	Alvarez (1989)
25	Yes	*1.48E − 08*[a]	Alvarez (1989)

[a]Representative value.

Table 111. Water solubility of permethrin.

Temp, °C	Column saturation	WS, ppm	Reference
25	No	2.20E – 01	Alvarez (1989)
25	No	1.30E – 01	Alvarez (1989)
25	Avg	1.75E – 01	
	CV, %	36.4	
20	Yes	3.50E – 03	Wollerton (1987)
20	Yes	6.10E – 03	Wollerton (1987)
20	Yes	6.90E – 03	Wollerton (1987)
20	Avg	5.50E – 03[a]	
	CV, %	32.3	

[a]Representative value.

ring. Temperature was lowered to 25 °C, and equilibration was established during continuous stirring of the solution with a magnetic stirrer. The first value in Table 111 is the solubility measured for *cis*-permethrin and the second that for *trans*-permethrin. These values were combined in the form of an average, which is also presented in the table. The average of the three solubilities reported by Wollerton is selected to represent permethrin because of the use of the generator column methodology, which is less likely to produce artificially high solubilities caused by formation of suspensions and not true solutions.

3. Henry's Law Constant. Henry's law constant for permethrin is calculated from Eq. 1, assuming a molecular weight of 391.3, vapor pressure of 1.5×10^{-8} mm Hg, and water solubility of 5.50×10^{-3} ppm to yield $K_h = 1.4 \times 10^{-6}$ atm m^3 mol^{-1}.

4. Octanol–Water Partition Coefficient. No reports were reviewed that describe the measurement of K_{ow} for permethrin. Wollerton (1987) cited a log P value for permethrin as 6.1 at 20 °C but did not supply any supporting details or data. Alvarez (1989) also referred to a value of log P = 6.1 but again did not cite any details. Tomlin (1994) used the same value, and so this value (referenced to Wollerton 1987) was selected to represent the K_{ow} for permethrin (Table 112). For purposes of comparison, CLog P, the value estimated by calcula-

Table 112. Permethrin octanol–water partition coefficient.

Best values	K_{ow}	Log P	CLog P	Reference
No	1.26E + 06	6.10[a]	6.9[b]	Wollerton (1987)

[a]Representative value.
[b]Calculated from molecular structure by Briggs (1999).

tion from the molecular structure of permethrin (Briggs 1999) is presented in the table and is noted as slightly greater than the measured value.

5. Bioconcentration Factor.   Bioconcentration of permethrin has been measured by Burgess (1989) in bluegill sunfish exposed continuously to permethrin in a flow-through proportional diluter apparatus for up to 28 d. Total ^{14}C concentrations in water and fish were measured periodically, and BCF cited in Table 113 is the average of the BCF values measured on days 21 and 28 in terms of total ^{14}C concentrations in the water and fish.

6. Soil Sorption.   Sorption of permethrin to soil was measured by Davis (1991), who added permethrin in acetonitrile to calcium chloride solutions (5% cosolvent) in Teflon centrifuges containing soil. Five soils and water with permethrin at four concentrations were equilibrated for 2 hr; samples were centrifuged, supernatants removed, and then replaced with fresh calcium chloride, and then reequilibrated another 48 hr for purposes of desorption measurement. Sorbed material was calculated by difference in ^{14}C content of the supernatant before and after equilibration with soil. TLC analysis of the supernatant from the highest adsorption treatment level indicated the following percentage of permethrin in the supernatant of each of the five soils used in the study: 97.6%, 88.9%, 91.8%, 87.1%, and 68.5%. Sorption K_d was not corrected for the presence of ^{14}C materials other than permethrin. Furthermore, the use of 5% cosolvent may have contributed to underestimation of K_d because of its impact on solution polarity.

The sorption was remeasured recently by Hand (2000), who equilibrated one concentration of permethrin for 3 hr with soil in dilute calcium chloride having only 50 µL acetonitrile per 20 mL. Supernatant was analyzed for total ^{14}C to determine distribution between aqueous phase and soil, and soil concentrations were calculated by difference between total ^{14}C applied and that remaining after equilibration. Desorption was measured by removal of adsorption supernatant, addition of fresh calcium chloride, and reequilibration for another 3 hr. After centrifugation and removal of supernatant, the supernatant was analyzed for total ^{14}C, and distribution between water and soil calculated as in the adsorption step.

Chromatographic analyses of soil extracts from desorption samples indicated

Table 113. Permethrin bioconcentration factor based on fish whole-body analysis.

Type of expt	Best values	BCF	Reference
Flow-through	Yes	558[a]	Burgess (1989)

[a]Representative value.

97.6%–98.9% of extracted material was permethrin; water-phase analyses did not extract all ^{14}C from the water into hexane, leaving 10%–30% of that in the aqueous phase. The material transferred to hexane was 95.5%–96.9% permethrin. K_d has not been corrected for the presence of ^{14}C material other than parent and so probably is underreported in this study, assuming ^{14}C not extracted from the water phase is other than permethrin. The work by Hand (2000) is selected as representative of the soil sorption of permethrin because it does not contain the cosolvent artifact present in the Davis dataset.

Permethrin adsorption K_d and K_{oc} from this study are listed in Table 114, and desorption coefficients in Table 115. Freundlich sorption data are shown in Table 116 only for the dataset generated by Davis (1991), because the work of Hand (2000) used a single concentration. As a result, the Freundlich data in Table 116 are discounted and not included in the summary chapter of this report because of the cosolvent artifact in the Davis study.

B. Abiotic Chemical Properties

1. Hydrolysis.   The hydrolysis of permethrin was measured by Allsup (1976) who added permethrin to sterile buffer solutions in glass-stoppered flasks and incubated them at 25 °C in the dark up to 28 d. Results from nonlinear and linear first-order analyses of datasets from this study are cited in Table 117. The first value at pH 9 in Table 117 is the rate constant for hydrolysis of the *trans* isomer of permethrin, the second value for *cis*-permethrin, and the third is the constant generated from the combined datasets of *cis* and *trans* isomers. The average of these three values is selected to represent the permethrin hydrolysis rate at pH 9.

2. Photolysis in Water.   Amos and Donelan (1987) reported the photolysis of permethrin in water. Permethrin ^{14}C-labeled in its cyclopropane or methylene group was added aseptically in 150 µL acetonitrile to 15 mL sterile pH 5 buffer in 20-mL glass tubes, and irradiated in temperature-controlled glass troughs filled with water. The troughs were made opaque so that only light from a Heraeus Hanau NXe 4500 xenon arc lamp, filtered to simulate Florida autumn sun, reached the samples through quartz lids over the troughs. Irradiation was up to the equivalent of 33 d under the simulated Florida autumn sunshine. Dark controls wrapped in aluminum foil were included, and analyses at 32 d indicated no degradation of permethrin had occurred in the absence of light. Correction of irradiated samples for dark reaction thus was not needed. Datasets from both labels were combined to develop the nonlinear and linear first-order kinetic data presented in Table 118.

3. Photolysis on Soil.   Photolysis of permethrin on soil was studied by Brown and Leahey (1987), who applied cyclopropane or methylene-labeled ^{14}C-permethrin dissolved in small quantities of hexane to the surface of air-dry soil on stainless steel plates. The plates were housed in a controlled-temperature cham-

Table 114. Permethrin soil adsorption partition coefficients.

Expt rating	Soil conc, ppm	Soil texture	K_d	K_{oc}	Reference
3	1	S	95.9	41,700	Davis (1991)
3	1	SL	176	16,900	Davis (1991)
3	1	SL	187	18,000	Davis (1991)
3	1	SL	294	18,700	Davis (1991)
3	1	CL	207	17,300	Davis (1991)
3	1	CL	208	17,300	Davis (1991)
3	1	CL	221	18,400	Davis (1991)
3	1	S	97	42,200	Davis (1991)
3	1	SL	181	17,400	Davis (1991)
3	1	S	97.4	42,300	Davis (1991)
3	1	SiL	206	16,900	Davis (1991)
3	1	SiL	203	16,600	Davis (1991)
3	1	SL	260	16,600	Davis (1991)
3	1	SiL	200	16,400	Davis (1991)
3	1	SL	308	19,600	Davis (1991)
3	3	CL	261	21,800	Davis (1991)
3	3	CL	238	19,800	Davis (1991)
3	3	SiL	243	19,900	Davis (1991)
3	3	SL	226	21,700	Davis (1991)
3	3	SL	212	20,400	Davis (1991)
3	3	S	143	62,200	Davis (1991)
3	3	CL	239	19,900	Davis (1991)
3	3	SiL	245	20,100	Davis (1991)
3	3	SiL	246	20,200	Davis (1991)
3	3	S	120	52,200	Davis (1991)
3	3	S	130	56,500	Davis (1991)
3	3	SL	219	21,100	Davis (1991)
3	4	SL	351	22,400	Davis (1991)
3	4	SL	353	22,500	Davis (1991)
3	4	SL	325	20,700	Davis (1991)
3	6	SL	235	22,600	Davis (1991)
3	6	SL	226	21,700	Davis (1991)
3	6	SL	235	22,600	Davis (1991)
3	6	S	167	72,600	Davis (1991)
3	6	S	164	71,300	Davis (1991)
3	6	S	131	57,000	Davis (1991)
3	6	SiL	239	19,600	Davis (1991)
3	6	SL	465	29,600	Davis (1991)
3	6	SL	468	29,800	Davis (1991)
3	6	SL	491	31,300	Davis (1991)
3	6	SiL	247	20,200	Davis (1991)
3	6	CL	264	22,000	Davis (1991)
3	6	SiL	246	20,200	Davis (1991)
3	6	CL	262	21,800	Davis (1991)

Table 114. (Continued).

Expt rating	Soil conc, ppm	Soil texture	K_d	K_{oc}	Reference
3	6	CL	268	22,300	Davis (1991)
6	8	SL	2,800	230,000	Hand (2000)
6	8	SL	2,400	200,000	Hand (2000)
6	8	SL	2,800	260,000	Hand (2000)
6	8	SL	3,100	280,000	Hand (2000)
6	8	SL	1,600	550,000	Hand (2000)
6	8	SL	1,500	520,000	Hand (2000)
6	8	SL	1,400	480,000	Hand (2000)
6	8	SL	2,700	250,000	Hand (2000)
6	8	LS	2,200	130,000	Hand (2000)
6	8	SL	2,000	170,000	Hand (2000)
6	8	LS	2,300	140,000	Hand (2000)
6	8	SL	2,400	200,000	Hand (2000)
6	8	SL	1,500	520,000	Hand (2000)
6	8	SL	3,000	270,000	Hand (2000)
6	8	LS	1,800	110,000	Hand (2000)
6	8	LS	2,100	120,000	Hand (2000)
3	10	S	176	76,500	Davis (1991)
3	10	S	182	79,100	Davis (1991)
3	10	S	182	79,100	Davis (1991)
3	10	CL	271	22,600	Davis (1991)
3	10	CL	262	21,800	Davis (1991)
3	10	SL	496	31,600	Davis (1991)
3	10	SL	240	23,100	Davis (1991)
3	10	SL	498	31,700	Davis (1991)
3	10	SiL	253	20,700	Davis (1991)
3	10	SiL	255	20,900	Davis (1991)
3	10	CL	254	21,200	Davis (1991)
3	10	SL	238	22,900	Davis (1991)
3	10	SL	234	22,500	Davis (1991)
3	10	SiL	243	19,900	Davis (1991)
3	10	SL	505	32,200	Davis (1991)
All ratings	Avg, <0.5				
	CV, %				
	Avg, 0.5 to <1				
	CV, %				
	Avg, 1–5		216	25,300	
	CV, %		32.1	51.4	
	Avg, >5		956	118,000	
	CV, %		104	125	
	Avg, All Conc		664	81,600	
	CV, %		128	152	

Table 114. (Continued).

Expt rating	Soil conc, ppm	Soil texture	K_d	K_{oc}	Reference
>4	Avg, <0.5				
	CV, %				
	Avg, 0.5 to <1				
	CV, %				
	Avg, 1–5				
	CV, %				
	Avg, >5		2,230[a]	277,000[a]	
	CV, %		25.1	55.4	
	Avg, All Conc		2,230	277,000	
	CV, %		25.1	55.4	

[a]Representative value.

ber and irradiated up to the equivalent of 33 d with artificial light from a Hanau NXe 4500 xenon arc lamp filtered to simulate Florida autumn sunlight. Dark controls analyzed at 30 d, included to correct for dark reaction, indicated permethrin was stable during the 30-d period, making a correction for dark reaction unnecessary. Datasets from the two ^{14}C-labeled forms of permethrin were combined to generate the kinetics information summarized in Table 119.

C. Biotic Chemical Properties

1. Aerobic Soil Degradation. Aerobic soil degradation of permethrin was studied by Williams and Brown (1979) and by Hawkins et al. (1991a, 1992). Williams and Brown added permethrin in benzene to a 10% portion of the total soil, evaporated the solvent, and mixed treated and untreated portions together. Aliquots were placed in pots and incubated at 20 °C for 15 hr followed by 10 °C for 9 hr up to 56 d. Overall incubation temperature was estimated to be equivalent to 16 °C.

Hawkins et al. (1991a) added permethrin in 100 µL acetonitrile or methanol to 60 g soil in dark continuous air flow-through chambers at 25 °C. Incubation lengths were up to 32 d. Using the same methodology in a second experiment (Hawkins et al. 1992) they expanded the periods of incubation from the initial 32 d to as long as 365 d, using two forms of permethrin labeled with ^{14}C in different structural positions. Nonlinear and linear first-order analyses of datasets from all three experiments are summarized in Table 120.

2. Anaerobic Soil Degradation. Hawkins et al. (1991b) utilized an anaerobic soil experimental design to study the degradation of permethrin under anaerobic conditions (Table 121). They applied permethrin in 100 µL acetonitrile with no additional mixing to the surface of 60 g soil in beakers housed in glass continu-

Table 115. Permethrin soil desorption partition coefficients.

Expt rating	Soil conc, ppm	Soil texture	K_{dd}	K_{ocd}	Reference
3	0.6	S	278	121,000	Davis (1991)
3	0.6	S	287	125,000	Davis (1991)
3	0.6	S	276	120,000	Davis (1991)
3	0.8	SL	74.3	7,140	Davis (1991)
3	0.9	SL	100	9,620	Davis (1991)
3	0.9	SL	96.4	9,270	Davis (1991)
3	1	SiL	56.8	4,660	Davis (1991)
3	1	CL	212	17,700	Davis (1991)
3	1	CL	205	17,100	Davis (1991)
3	1	SL	2,270	145,000	Davis (1991)
3	1	CL	189	15,800	Davis (1991)
3	1	SiL	59	4,840	Davis (1991)
3	1	SL	1,680	107,000	Davis (1991)
3	1	SL	2,270	144,000	Davis (1991)
3	1	SiL	55.8	4,570	Davis (1991)
3	2	S	237	103,000	Davis (1991)
3	2	S	280	122,000	Davis (1991)
3	3	CL	237	19,800	Davis (1991)
3	3	SL	120	11,500	Davis (1991)
3	3	SL	115	11,100	Davis (1991)
3	3	SL	108	10,400	Davis (1991)
3	3	S	286	124,000	Davis (1991)
3	3	S	240	104,000	Davis (1991)
3	3	CL	212	17,700	Davis (1991)
3	3	S	282	123,000	Davis (1991)
3	3	CL	201	16,800	Davis (1991)
3	3	SiL	73.5	6,020	Davis (1991)
3	3	SiL	77.2	6,330	Davis (1991)
3	3	SiL	76.1	3,780	Davis (1991)
3	4	SL	1,990	127,000	Davis (1991)
3	4	SL	2,030	129,000	Davis (1991)
3	4	SL	2,000	127,000	Davis (1991)
3	5	S	271	118,000	Davis (1991)
3	5	S	328	143,000	Davis (1991)
3	5	SiL	98.6	8,080	Davis (1991)
3	5	S	276	120,000	Davis (1991)
3	5	SiL	96.5	7,910	Davis (1991)
3	6	CL	285	23,800	Davis (1991)
3	6	SL	123	11,800	Davis (1991)
3	6	SL	2,690	171,000	Davis (1991)
3	6	SL	134	12,900	Davis (1991)
3	6	SL	136	13,100	Davis (1991)
3	6	SiL	149	9,340	Davis (1991)
3	6	CL	277	23,100	Davis (1991)
3	6	CL	268	22,300	Davis (1991)

Table 115. (Continued).

Expt rating	Soil conc, ppm	Soil texture	K_{dd}	K_{ocd}	Reference
3	7	SL	2,720	173,000	Davis (1991)
3	7	SL	2,650	169,000	Davis (1991)
6	7	SL	1,700	590,000	Hand (2000)
6	7	LS	2,100	120,000	Hand (2000)
6	7	LS	2,000	120,000	Hand (2000)
6	7	SL	1,800	620,000	Hand (2000)
6	8	SL	2,700	230,000	Hand (2000)
6	8	SL	3,900	360,000	Hand (2000)
6	8	SL	3,800	350,000	Hand (2000)
6	8	SL	2,900	240,000	Hand (2000)
3	8	SiL	114	9,340	Davis (1991)
3	8	SL	205	19,700	Davis (1991)
3	9	SiL	111	9,100	Davis (1991)
3	9	CL	318	26,500	Davis (1991)
3	9	SiL	117	9,590	Davis (1991)
3	9	SL	196	18,800	Davis (1991)
3	9	SL	180	17,300	Davis (1991)
3	9	CL	283	23,600	Davis (1991)
3	10	CL	310	25,800	Davis (1991)
3	11	SL	2,930	186,000	Davis (1991)
3	11	SL	2,910	185,000	Davis (1991)
3	11	SL	2,830	180,000	Davis (1991)
All ratings	Avg, <0.5				
	CV, %				
	Avg, 0.5 to <1				
	CV, %				
	Avg, 1–5		599	58,600	
	CV, %		136	97.8	
	Avg, >5		1,200	125,000	
	CV, %		109	125	
	Avg, All Conc		874	93,800	
	CV, %		128	131	
>4	Avg, <0.5				
	CV, %				
	Avg, 0.5 to <1				
	CV, %				
	Avg, 1–5				
	CV, %				
	Avg, >5		2,610	329,000	
	CV, %		33.3	58.5	
	Avg, All Conc		2,610	329,000	
	CV, %		33.3	58.5	

Table 116. Freundlich sorption data for permethrin.[a]

Expt rating	Soil texture	K_{df}	K_{ocf}	$1/n$	Reference
3	S	446	194,000	1.32	Davis (1991)
3	SL	355	34,100	1.12	Davis (1991)
3	CL	378	31,500	1.1	Davis (1991)
3	SiL	344	28,200	1.09	Davis (1991)
3	SL	1,520	96,600	1.29	Davis (1991)
All ratings	Avg	609	76,900	1.18	
	CV, %	84.0	92.8	9.4	
>4	Avg				
	CV, %				

[a]Not used in summary chapter because of cosolvent artifact.

Table 117. Hydrolysis of permethrin in buffered water.

Temp, °C	pH	Nonlinear k^a	Nonlinear Half-life, d[a]	Linear k^a	Linear Half-life, d[a]	Reference
25	3	$0.00E+00^b$	0	$0.00E+00$	0	Allsup (1976)
25	6	$0.00E+00^b$	0	$0.00E+00$	0	Allsup (1976)
25	9	3.73E−03	186	3.58E−03	194	Allsup (1976)
25	9	2.03E−03	341	1.99E−03	348	Allsup (1976)
25	9	2.83E−03	245	2.77E−03	250	Allsup (1976)
25	Avg	$2.86E−03^b$		2.78E−03		
	CV, %	29.7		28.6		

[a]An entry of 0.00E + 00 means k was measured but permethrin is stable; a blank k or half-life value means no measurement.
[b]Representative value.

Table 118. Photolysis of permethrin in water.

Natural sun	Buffer medium	Nonlinear k	Nonlinear Half-life, d	Linear k	Linear Half-life, d	Reference
No	Yes	$6.31E−03^a$	110	6.73E−03	103	Amos and Donelan (1987)

[a]Representative value.

Table 119. Photolysis of permethrin on soil.

Natural sun	Buffer medium	Nonlinear		Linear		Reference
		k	Half-life, d	k	Half-life, d	
No	Yes	*6.66E − 03*[a]	104	6.51E − 03	106	Brown and Leahey (1987)

[a]Representative value.

ous gas flow-through towers and incubated the samples for 30 d aerobically in the dark. Then samples were flooded with water, continuous stream of gas changed from air to nitrogen, and incubated anaerobically for up to an additional 60 d. Data cited in Table 121 are from composite datasets from two ^{14}C labels; experimental rating is set to less than 5 because of the high rate of application (13 ppm) with no additional mixing.

3. Aerobic Aquatic Degradation.   No references were reviewed that considered the degradation of permethrin in aerobic aquatic systems.

4. Anaerobic Aquatic Degradation.   No anaerobic experiments were reviewed that utilized the anaerobic aquatic experimental design for permethrin.

Summary

The physical and chemical properties of the pyrethroids bifenthrin, cyfluthrin, cypermethrin (also zetacypermethrin), deltamethrin, esfenvalerate (also fenvalerate), fenpropathrin, lambda-cyhalothrin (also cyhalothrin), permethrin, and tralomethrin have been reviewed and summarized in this paper. Physical properties included molecular weight, octanol–water partition coefficient, vapor pressure, water solubility, Henry's law constant, fish biocencentration factor, and soil sorption, desorption, and Freundlich coefficients. Chemical properties included rates of degradation in water as a result of hydrolysis, photodecomposition, aerobic or anaerobic degradation by microorganisms in the absence of light, and also rates of degradation in soil incubated under aerobic or anaerobic conditions.

Collectively, the pyrethroids display a highly nonpolar nature of low water solubility, low volatility, high octanol–water partition coefficients, and have high affinity for soil and sediment particulate matter. Pyrethroids have low mobility in soil and are sorbed strongly to the sediments of natural water systems. Although attracted to living organisms because of their nonpolar nature, their capability to bioconcentrate is mitigated by their metabolism and subsequent elimination by the organisms. In fish, bioconcentration factors (BCF) ranged from 360 and 6000.

Pyrethroids in water solution tend to be stable at acid and neutral pH but

Table 120. Permethrin degradation in aerobic soil.

Expt rating	Temp, °C	Soil conc, ppm	Soil texture	Nonlinear		Linear		Reference
				k	Half-life, d	k	Half-life, d	
5	16	1	CL	3.51E – 02	19.7	3.99E – 02	17.4	Williams and Brown (1979)
5	16	1	SiL	5.87E – 02	11.8	4.34E – 02	16	Williams and Brown (1979)
8	25	1	SL	6.00E – 02	11.6	5.75E – 02	12.1	Hawkins et al. (1991a)
8	25	1	SL	4.24E – 02	16.3	3.68E – 02	18.8	Hawkins et al. (1991a)
8	25	1	SL	3.05E – 02	22.7	2.96E – 02	23.4	Hawkins et al. (1991a)
7	25	1.6	SL	1.94E – 02	35.7	9.56E – 03	72.5	Hawkins et al. (1992)
8	25	13	SL	8.14E – 03	85.2	8.01E – 03	86.5	Hawkins et al. (1991a)
8	25	13	SL	6.14E – 03	113	6.15E – 03	113	Hawkins et al. (1991a)
All ratings		Avg, <0.5 CV, % Avg + CL						
		Avg, 0.5–1.5 CV, % Avg + CL			16.4 29.6 19.8		17.5 23.5 20.4	
		Avg, 1.5–10 CV, % Avg + CL			35.7 107		72.5 218	
		Avg, >10 CV, % Avg + CL			99.1 19.8 142		99.8 18.8 141	
		Avg, All Conc CV, % Avg + CL			39.5 96.9 58.7		45.0 87.9 64.7	
>4		Avg, All Conc CV, % Avg + CL			39.5[a] 96.9 58.7[a]		45.0 87.9 64.7	

[a]Representative value.

become increasingly susceptible to hydrolysis at pH values beyond neutral. Exceptions at higher pH are bifenthrin (stable), esfenvalerate (stable), and permethrin (half-life, 240 d).

Pyrethroids vary in susceptibility to sunlight. Cyfluthrin and tralomethrin in water had half-lives of 0.67 and 2.5 d; lambda-cyhalothrin, esfenvalerate, deltamethrin, permethrin, and cypermethrin were intermediate with a range of 17–110 d; and bifenthrin and fenpropathrin showed the least susceptibility with half-lives of 400 and 600 d, respectively. Pyrethroids on soil can also undergo photolysis, often at rates similar to that in water. Half-lives ranged from 5 to 170 d.

Table 121. Permethrin degradation generated by an anaerobic soil experimental design.

Expt rating	Temp, °C	Soil conc, ppm	Soil texture	Nonlinear		Linear		Reference
				k	Half-life, d	k	Half-life, d	
3	25	13	SL	3.52E − 03	197	3.42E − 03	203	Hawkins et al. (1991b)
All ratings		Avg, <1.5 CV, % Avg + CL						
		Avg, >1.5 CV, % Avg + CL			197[a] 591[a]		203 609	
		Avg, All Conc CV, % Avg + CL			197 591		203 609	
>4		Avg, All Conc CV, % Avg + CL						

[a]Representative value.

Pyrethroids are degradable in soils with half-lives ranging from 3 to 96 d aerobically, and 5 to 430 d anaerobically. For those pyrethroids studied in water (cypermethrin, deltamethrin, esfenvalerate, fenpropathrin, and lambda-cyhalothrin), aerobic and anaerobic degradation often continued at rates similar to that displayed in soil.

Acknowledgments

The author thanks the Ecotoxicity Technical Committee of the Pyrethroid Working Group for providing the funding and resources needed to prepare this review and manuscript. A special thank-you is extended to Michael Dobbs, Chairman of the Ecotoxicity Committee, for his constant support and help throughout the project, and to Richard Allen, Zenna Burke, Paul Hendley, Nancy Hilton, Ray Layton, Steve Maund, Jacqui Warinton, and other members of the Ecotoxicity Technical Committee for their very helpful comments and suggestions during the development and preparation of this review.

References

Agrevo a (unknown date) RU 25474: study on the hydrolysis. AgrEvo report A73713.
Agrevo b (unknown date) Untitled. AgrEvo report A74007; EPA MRID 099741.

Allsup TL (1976) Hydrolysis of FMC 33297 insecticide. FMC report W-0103. EPA,

Alvarez M (1989) Permethrin: physical properties. FMC report P-2242, EPA MRID 42109801.

Alvarez M (1991a) Physical properties of FMC 56701. FMC report P-2595, EPA MRID # 41968203. EPA,

Alvarez M (1991b) Physical properties of cypermethrin. FMC report P-2594, EPA MRID 41887003. EPA,

Alvarez M (1995) Zetacypermethrin: octanol-water partition coefficient. FMC report P-3040.

Amos R, Donelan RB (1987) Permethrin: photolysis in sterile water at pH 5. Zeneca report RJ0577B; EPA MRID 40242801. U.S. Environmental Protection Agency.

Baldwin MK, Lad D (1978) The accumulation and elimination of WL 43467 by the rainbow trout (*Salmo gairdneri*). Zeneca report Shell TLGR.0041.78.

Bennett D (1981) The accumulation, distribution and elimination of RIPCORD by rainbow trout using a continuous-flow procedure. Zeneca report; Shell report SBGR. 81.026. Shell,

Bharti H, Bewick DW, White RD (1985) PP563 and PP321: degradation in soil. Zeneca report RJ 0382B.

Bixler TA, Gross E, Willow E (1983) FMC 54800 aerobic soil degradation. FMC report P-0712; EPA MRID 532540.

Bowman B, Carpenter M (1987) Determination of photodegradation of ^{14}C-deltamethrin in aqueous solution. AgrEvo report A41919; EPA MRID 40254101.

Briggs G (1999) Personal communication. AgrEvo,

Brown PM, Leahey JP (1987) Permethrin: photolysis on a soil surface. Zeneca report RJ0581B; EPA MRID 40190101.

Burgess D (1989) Uptake, depuration and bioconcentration of ^{14}C-permethrin by bluegill sunfish (*Lepomis macrochirus*). FMC report PC-0117; EPA MRID 41300401.

Burhenne J (1996) Adsorption/desorption of cyfluthrin on soils. Bayer report 107397. Bayer,

Carlisle JC, Roney DJ (1984) Bioconcentration of cyfluthrin (BAYTHROID) by bluegill sunfish. Bayer report 86215.

Castle S, Shepler K, Ruzo LO (1990) Photodegradation of [^{14}C]esfenvalerate in/on soil surface by natural sunlight. DuPont report AMR-1798–90. DuPont,

Cheng HM (1986) Characterization of ^{14}C residues in bluegill sunfish treated with ^{14}C-fenpropathrin (revised). Valent report 9109227.

Chopade HM (1986) Photodecomposition of [^{14}C] BAYTHROID on soil. Bayer report 88981.

Christensen KP (1993) Deltamethrin: determination of the sorption and desorption properties. AgrEvo report A73876; EPA MRID 42976501.

Clifton JL (1992) Environmental fate studies: hydrolysis studies of cypermethrin in aqueous buffered solutions. FMC report P-2771; EPA MRID 42620501.

Collis WMD, Leahey JP (1984) PP321: hydrolysis in water at pH 5, 7, and 9. Zeneca report RJ0338B.

Concha MA, Ruzo LO, Shepler K (1992a) Photodegradation of [^{14}C-acid] and [^{14}C-alcohol] fenpropathrin in/on soil by natural sunlight. Valent report 9200548; EP 4254 6402.

Concha MA, Shepler K, Ruzo LO (1992b) Hydrolysis of [^{14}C-acid] and [^{14}C-alcohol]fenpropathrin at pH 5, 7 and 9. Valent report 9200714; EPA 141320253931.

Cranor W (1990) Aerobic soil metabolism of [benzyl-^{14}C]-fenpropathrin. Valent report 9004803C; EPA MRID 42525902.

Curl EA, Leahey JP, Lloyd SJ (1984a) PP321: aqueous photolysis at pH 5. Zeneca report RJ0362B.

Curl EA, Leahey JP, Lloyd S (1984b) PP321: photodegradation on a soil surface. Zeneca report RJ0358B.

Daly D (1989) Soil/sediment adsorption-desorption with ^{14}C-tralomethrin. AgrEvo report A72678.

Daly D, Williams M (1990) Anaerobic soil metabolism of ^{14}C-fenpropathrin. Valent report 9200549; EPA MRID 42546403.

Davis ML (1991) Sorption/desorption of ^{14}C-permethrin on soils by the batch equilibrium method. FMC report PC-0156; EPA MRID 41868001.

Devaux PH, Bolla P (1984) Photodegradation of tralomethrin in water. AgrEvo report A72969.

Ehman A, Ingamells JM (1981) Photodegradation of SD 43775 on soil thin layers. DuPont report RIR-22–002-81.

Elmarakby SA (1998) Aerobic aquatic metabolism of ^{14}C-zeta-cypermethrin. FMC report P-3312.

Estigoy L, Ruzo LO, Shepler K (1991a) Photodegradation of [^{14}C-acid]- and [^{14}C-alcohol]-cypermethrin in/on soil by natural sunlight. FMC report PC-0159; EPA MRID 42129001.

Estigoy L, Ruzo LO, Shepler K (1991b) Photodegradation of [^{14}C-acid] and [^{14}C-alcohol] cypermethrin in buffered aqueous solution at pH 7 by natural sunlight. FMC report PC-0163; EPA MRID 42141501.

Fackler PH (1990) Deltamethrin: bioconcentration and elimination of ^{14}C residues by bluegill (Lepomis macrochirus). AgrEvo report A47117; EPA MRID 41651040.

Forbis AD (1985) Uptake, depuration and bioconcentration of [cyclopropyl-1–^{14}C] and benzyl-1–^{14}C] fenpropathrin by bluegill sunfish (Lepomis macrochirus). Valent report 9109227; EPA MRID 161672.

Froelich LW (1983) Soil adsorption/desorption characteristics of FMC 54800. FMC report P-0797; EPA MRID 141203.

Froelich LW (1991) Soil mobility studies: adsorption/desorption studies of cypermethrin. FMC report P-2658; EPA MRID 42129003.

Gaddamidi V, Bookhart SW III (1992) Anaerobic soil metabolism of esfenvalerate. DuPont report AMR 2075–91.

Gargot A (1994) Tralomethrin: determination of the octanol/water partition coefficient. AgrEvo report A74128.

Giroir LE, Stuerman L (1993) [^{14}C]Cypermethrin bioconcentration by bluegill sunfish (Lepomis macrochirus). FMC report PC-0189; EPA MRID 42868203.

Goggin U, Gentle W, Hamer MJ, Lane M CG (1996) Cypermethrin: adsorption and desorption properties in sediment. Zeneca report RC0004.

Grayson BT, Langner E, Wells D (1982) Comparison of two gas saturation methods for the determination of the vapour pressure of cypermethrin. Pestic Sci 13:552–556.

Grelet D (1990) Deltamethrin: active Ingredient. Summary of physical and chemical characteristics. AgrEvo report A70742; EPA MRID 41651003.

Gronberg RR (1984) Photodecomposition of [phenyl-UL-^{14}C] BAYTHROID in aqueous solution by sunlight. Bayer report 88598.

Gronberg RF (1987) Adsorption of BAYTHROID to sandy loam. Bayer report 94541.

Hall JS, Leahey JP (1983) Cyhalothrin: fate in river water. Zeneca report RJ 0320B.

Hand LH (2000) Permethrin: adsorption and desorption properties in four soils. Zeneca technical letter 00JH005/01 (interim report).

Hansch C, Leo A (1979) Substituent Constants for Correlation Analysis in Chemistry and Biology. Wiley, New York.

Harvey BR, Zinner CKJ, White RD, Hill IR (1981) Cypermethrin: degradation in soil in the laboratory. Zeneca report RJ 0162 B.

Hawkins B, Kirkpatrick D, Shaw D, Riseborough J (1991a) The effect of application rate and soil moisture content on the rate of degradation of ^{14}C-permethrin in aerobic sandy loam soil. FMC report HRC/ISN 247/91296; EPA MRID 41970602.

Hawkins B, Kirkpatrick D, Shaw D, Riseborough J (1991b) The metabolism of ^{14}C-permethrin in sandy loam soil under anaerobic conditions. FMC report HRC/ISN 236/91107; EPA MRID 41970601.

Hawkins DR, Kirkpatrick D, Shaw D, Nicholson J (1992) The aerobic soil metabolism of ^{14}C-permethrin. FMC report HRC/ISN 251/911499; EPA MRID 42410002.

Hellpointner E (1990) Determination of the quantum yield and assessment of the environmental half-life of the direct photodegradation of cyfluthrin in water. Bayer report 103228.

Herbst RM (1983a) Water solubility of FMC 54800. FMC report P-0699; EPA MRID 251725.

Herbst RM (1983b) Octanol water partition coefficient of FMC 54800. FMC report P-068; EPA MRID 251725.

Herbst RM (1983c) Hydrolysis of FMC 54800. FMC report P-0701; EPA MRID 132539.

Itoh K, Kodaka R, Kumada K, Nambu K, Kato T (1995) Aerobic soil metabolism of esfenvalerate and fenvalerate in European soils. DuPont report LLM-50-0039.

Jalai-Araghi K, Ruzo LO, Shepler K (1992) Photodegradation of [^{14}C-acid] and [^{14}C-alcohol] fenpropathrin in a buffered aqueous solution at pH 5 by natural sunlight. Valent report 9200547; EPA 42546402.

Kaman RA (1994) An anaerobic aquatic soil metabolism study with [cyclopropyl-1–^{14}C] fenpropathrin. Valent report 9600477; EPA MRID 44370004.

Kaman RA (1995) An anaerobic aquatic soil metabolism study with [phenoxyphenyl-^{14}C]-fenpropathrin. Valent report 9600477; EPA MRID 44370003.

Kaufman DD, Kayser AJ, Russell BA (date unknown a) Degradation of the synthetic pyrethroid insecticide RU-25474 in soil. AgrEvo report A73235.

Kaufman DD, Kayser AJ, Doyle EH, Munitz T (date unknown b) Degradation of ^{14}C-methylene- and ^{14}C-gem-methyl-tralomethrin in three soils. AgrEvo report A72967.

Kaufman DD, Kayser AJ, Doyle EH, Munitz T (date unknown c) Anaerobic degradation of ^{14}C-gem-methyl and ^{14}C-methylene tralomethrin in flooded soil. AgrEvo report A72965.

Kaufman DD, Kayser AJ, Russell B, Barnett EA (1990a) The effect of soil temperature on the degradation of ^{14}C-cyano-decamethrin in soil. AgrEvo report A71051; EPA MRID 41677405.

Kaufman DD, Kayser AJ, Russell B, Barnett EA (1990b) Degradation of ^{14}C-phenoxy- and ^{14}C-cyano-decamethrin in soil. AgrEvo report A71064; EPA MRID 41677404.

Krohn J (1983a) Properties of pesticides in water. Bayer report 85986.

Krohn J (1983b) Water solubility of cyfluthrin. Bayer report 86626.

Krohn J (1987a) Octanol water partition coefficient of cyfluthrin pure active ingredient. Bayer report 94669.

Krohn J (1987b) Water solubility of cyfluthrin pure active ingredient. Bayer report 94668.

Krohn J (1988) Water solubility of cyfluthrin K+L [FCR 4545]. Bayer report 98320.

Lee DY (1992) Adsorption and desorption of fenpropathrin to soils. Valent report 9200731; EPA 42584101.

Lee P (1979) Twelve months aerobic soil metabolism of ^{14}C-chlorophenyl-SD43775. DuPont report AMR-1578-89, Appendix I.

Lee P, Stackhouse SC (1979) Comparative aerobic metabolism of ^{14}C-chlorophenyl-SD 43775 in sterilized and nonsterilized Hanford sandy loam soil. DuPont report AMR-1578-89, Appendix IV.

Lee P, Stearns S, Powell W (1985) Comparative aerobic soil metabolism of SD 43775 (racemic) and SD 47443 (A-alpha). DuPont report AMR-1578-89, Appendix V; EPA MRID 00146578.

Lee PW (1985) Fate of fenvalerate (Pydrin insecticide) in the soil environment. DuPont report AMR-1578-89, Appendix VII.

Lee PW (1988) Hydrolysis of [chlorophenyl-^{14}C] DPX-GB800 in buffer solutions of pH 5, 7, and 9. DuPont report AMR-1185-88.

Lee PW (1990) Bioaccumulation of fenvalerate in fish. DuPont report AMR-1827-90.

Lewis CJ (1995) (^{14}C)-Esfenvalerate: biodegradation in natural water-sediment systems at 10 °C. DuPont report LLM-51-0040.

Lorence PJ (1991) Fenpropathrin: determination of vapor pressure. Valent report 9200342.

Lorence PJ (1996) Fenpropathrin (S-3206): water solubility. Valent report 9600092; EPA MRID 44370001.

Lucas T (1998) Final report. [^{14}C]-zeta cypermethrin: aerobic aquatic degradation in two water/sediment systems. FMC report PC-0298.

Marriott SH, Duley J, Hand L (1998) Lambda-cyhalothrin: degradation in water–sediment systems under laboratory conditions. Zeneca report RJ 2640B.

Mikami N, Takahashi N, Hayashi K, Miyamoto J (1980) Photodegradation of fenvalerate (Sumicidin) in water and on soil surface. J Pestic Sci 5:225–236.

Mikami N, Sakata S, Yamada H, Miyamoto J (1983) Degradation of fenpropathrin in soil. Valent report 9300185; EPA 126823249936.

Mikami N, Sakata S, Yamada H (1984a) Further studies on degradation of the pyrethroid insecticide fenvalerate in soils. DuPont report AMR-1578-89, Appendix VI.

Mikami N, Sakata S, Yamada H (1984b) The response to EPA's requirement concerning soil studies of fenpropathrin (Danitol): data on degradation in soils under aerobic conditions at shorter time points. Valent report 9102225; EPA 137768252780.

Moffatt F (1994a) Cypermethrin: quantum yield and environmental half-life for direct phototransformation in aqueous solution. Zeneca report RJ1667B.

Moffatt F (1994b) Lambda-cyhalothrin: environmental half-life and quantum yield for direct phototransformation in aqueous solution. Zeneca report RJ1617B.

Muller K, Goggin U, Lane MCG (1996) Lambda-cyhalothrin: adsorption and desorption in soil and sediment. Zeneca report RJ 1913B

Muttzall PI (1993) Water/sediment biodegradation of [benzyl-^{14}C] deltamethrin. AgrEvo report A50953.

Ohkawa H, Nambu K, Inui H (1978) Metabolic fate of fenvalerate (Sumicidin) in soil and by soil microorganisms. DuPont report AMR-1578-89, Appendix III.

Parker S, Leahey JP (1986) PP321: photodegradation on a soil surface. Zeneca report RJ0537B.

Pepin M, Gargot A (1988) Tralomethrin solubility study in water. AgrEvo report A73322.

Pionke HB, DeAngelis RJ (1980) Method for distributing pesticide loss in field runoff between the solution and adsorbed phase. In: Knisel WG (ed) CREAMS: a field scale model for chemicals, runoff, and erosion from agricultural management systems. Report 26, Chapter 19. U.S. Department of Agriculture Conservation, Washington, DC.

Potter JC, Arnold DL (1980) Twelve-month aerobic soil metabolism of ^{14}C-phenoxyphenyl SD 43775. DuPont report AMR-1578-89, Appendix II.

Priestley DB, Leahey JP (1988) PP321: aqueous photolysis at pH 5. Zeneca report RJ0605B.

Puhl RJ, Hurley JB, Dime RA (1983) Photodecomposition of BAYTHROID ^{14}C in aqueous solution and on soil. Bayer report 86182.

Ramsey AA (1991a) Environmental fate studies: aerobic soil metabolism of cypermethrin in a sandy loam soil. FMC report P-2616; EPA MRID 42156601.

Ramsey AA (1991b) Environmental fate studies: anaerobic soil metabolism of cypermethrin in a sandy loam soil. FMC report P-2617; EPA MRID 42156602.

Ramsey AA (1998) Anaerobic aquatic metabolism of ^{14}C-zeta-cypermethrin. FMC report P-3329.

Rapley JH, Arnold DJ, Vincent J (1981) Cypermethrin: degradation in river and pond waters and sediments. Zeneca report RJ 0175B.

Rekker R (1977) The Hydrophobic Fragmental Constant. Elsevier, Amsterdam.

Reynolds JL (1984) Aerobic soil metabolism of FMC 54800: fate of acid cyclopropyl ring)-^{14}C FMC 54800 and metabolite characterization. FMC report P-0872; EPA MRID 141202.

Reynolds JL (1986) Metabolism of acid (cyclopropyl ring)-^{14}C and alcohol (phenyl ring)-^{14}C FMC 54800 in soil under anaerobic conditions. FMC report P-1338; EPA MRID 264642.

Roberts TR, Standen ME (1976) The degradation of the insecticide WL 41706 in soil under laboratory conditions. Valent report 9300185; EPA 130868126822.

Saito S, Itoh K (1992) Water solubility of fenpropathrin. Valent report 9200341; EPA MRID 424998-01

Sandie FE (1983) Hydrolysis of BAYTHROID in sterile, aqueous buffered solutions. Bayer report 86051.

Schocken MJ (1993) Deltamethrin: bioconcentration exposure with bluegill sunfish (*Lepomis macrochirus*) and identification of resulting metabolites. AgrEvo report A70918; EPA MRID 43072701.

Sewekow B (1981) Determination of vapor pressure of cyfluthrin. Bayer report 86627.

Shelby DJ (1996) Report amendment to: Adsorption and desorption of fenpropathrin to soils MRID 42584101. Valent report 9600478; EPA MRID 44370002.

Smith AM (1990a) Determination of aqueous hydrolysis rate constant and half-life of deltamethrin. AgrEvo report A43049; EPA MRID 41651038.

Smith AM (1990b) Determination of the adsorption and desorption coefficients of deltamethrin. AgrEvo report A47159; EPA MRID 41651039.

Stevenson IE (1987) Photodegradation of [chlorophenyl (U)-^{14}C]DPX-GB800 in water at pH 5. DuPont report AMR-868-87.

Suprenant DC (1986) Accumulation and elimination of ^{14}C-residues by bluegill (*Lepomis machrochirus*) exposed to ^{14}C-FMC 54800. FMC report PC-0038; EPA MRID 264642.

Swaine H, Hayward GJ (1979) Cypermethrin: laboratory degradation on two standard soils, Part I. Zeneca report RJ 0115B.

Takahashi N, Mikami N, Yamada H, Miyamoto J (1983a) Photodegradation of fenpropathrin in water, and on soil and plant foliage. Valent report 9200622.

Takahashi N, Mikami N, Yamada H, Miyamoto J (1983b) Hydrolysis of fenpropathrin in aqueous media. Valent report 9200620; EPA 131438251415.

Takimoto Y, Ohshima M, Matsuda T, Miyamoto J (1985) Accumulation and metabolism of [benzyl-1-^{14}C]-fenpropathrin in carp (*Cyprinus carpio*). Valent report 9109227; EPA MRID 153802.

Talbot TD, Mosier B (1987) Vapor pressure of BAYTHROID pure active ingredient. Bayer report 94330.

Tillier C (1993) RU 25474: determination of vapor pressure. AgrEvo report A74125.

Tomlin C (1994) The Pesticide Manual, 10th Ed. British Crop Protection Council and Royal Society of Chemistry, United Kingdom.

Valent (1983) Partition coefficient (*n*-octanol/water) of fenpropathrin. Valent report 9102150.

Wagner K, Neitzel H, Oehlmann L (1983) Decomposition of BAYTHROID in soil under aerobic and anaerobic conditions. Bayer report 86052.

Wang WW (1990a) Soil photolysis of ^{14}C-tralomethrin. AgrEvo report A72955.

Wang WW (1990b) Aerobic soil metabolism of ^{14}C-tralomethrin. AgrEvo report A72956.

Wang WW (1990c) Hydrolysis of ^{14}C-tralomethrin in water at pH 4, pH 5, pH 7, and pH 9. AgrEvo report A45108.

Wang WW (1990d) Anaerobic soil metabolism of ^{14}C-tralomethrin. AgrEvo report A72957.

Wang WW (1991a) Aqueous photolysis of ^{14}C-tralomethrin. AgrEvo report A72954.

Wang WW (1991b) Anaerobic metabolism of ^{14}C-deltamethrin. AgrEvo report A47918; EPA MRID 42114821.

Wang WW (1991c) Aerobic soil metabolism of ^{14}C deltamethrin. AgrEvo report A47917; EPA MRID 42114820.

Wang WW, Reynolds JL (1991a) Aqueous photolysis of ^{14}C-deltamethrin. AgrEvo report A47960; EPA MRID 42114818.

Wang WW, Reynolds JL (1991b) Soil photolysis of ^{14}C-deltamethrin. AgrEvo report A47919; EPA MRID 42114819.

Warren J (1984) Photodegradation of tralomethrin on soil surface exposed to artificial sunlight. AgrEvo report A72966.

Williams IH, Brown MJ (1979) Persistence of permethrin and WL 43775 in soil. J Agric Food Chem 27:130–132.

Wollerton C (1987) Permethrin water solubility and octanol-water partition coefficient. Zeneca summary, September 2, 1987.

Wollerton C, Husband R (1988a) PP321: water solubility, octanol-water partition coefficient, vapour pressure, and Henry's law constant. Zeneca report RJ 0699B.

Wollerton C, Husband R (1988b) Cypermethrin: water solubility, octanol-water partition coefficient and Henry's law constant. Zeneca report RJ0672B.

Wolt JD (1996) Personal communication to the FIFRA Exposure Model Validation Task Force.

Wu J, Gross EM, Gavin D (1986a) Photodegradation of FMC 54800 in aqueous solution. FMC report P-1349; EPA MRID 264642.

Wu J, Gross EM, Gavin D (1986b) Photodegradation of FMC 54800 in/on soil. FMC report P-1351; EPA MRID 264642.

Yamauchi F (1985) PP-563 Cyhalothrin: accumulation in fish (carp) in a flow-through water system. Zeneca report; MITES report 58-367.

Yoder SJ (1991a) Deltamethrin A.I. Determination of vapor pressure. AgrEvo report
    A47916; EPA MRID 42137502.
Yoder SJ (1991b) Deltamethrin A.I. Determination of octanol/water partition coefficient.
    AgrEvo report A47915; EPA MRID 42114802.
Yoshida H, Yoshimoto Y, Takase I (1984) Residual fate of cyfluthrin (FCR 1272) in
    soils under laboratory and field conditions. Bayer report 1197.

Manuscript received May 26; accepted June 21, 2001.

Index